ALSO BY JENNIFER JACQUET

Is Shame Necessary?: New Uses for an Old Tool

THE PLAYBOOK

THE PLAYBOOK

HOW TO DENY SCIENCE, SELL LIES, AND MAKE A KILLING IN THE CORPORATE WORLD

JENNIFER JACQUET

Pantheon Books, New York

 Published in the United States by Pantheon Books, a division of Penguin Random House LLC, New York, and distributed in Canada by Penguin Random House Canada Limited, Toronto.

Pantheon Books and colophon are registered trademarks of Penguin Random House LLC.

Library of Congress Cataloging-in-Publication Data
Name: Jacquet, Jennifer, author.
Title: The playbook: how to deny science, sell lies, and make a killing in the corporate world / Jennifer Jacquet.
Description: New York: Pantheon Books, 2022
Identifiers: LCCN 2021053111 (print) | LCCN 2021053112 (ebook) | ISBN 9781101871010 (hardcover) | ISBN 9781101871027 (ebook)
Subjects: LCSH: Communication in organizations. Denial (Psychology) | Defense mechanisms (Psychology)
Classification: LCC HD30.3 .J325 2022 (print) | LCC HD30.3 (ebook) | DDC 658.4/5—dc23/eng/20211213
LC record available at https://lccn.loc.gov/2021053111
LC ebook record available at https://lccn.loc.gov/2021053112

www.pantheonbooks.com

Jacket images: (smokestacks) ViewStock; (people) moodboard/Image Source; (clouds) EujarimPhotography/Moment; all Getty Images
Jacket design by Tyler Comrie

Printed in the United States of America
1st Printing

CONTENTS

DETAILED CONTENTS

3.
RECRUITING UNIVERSITY EXPERTS

- So you want to work for industry?
- What industry looks for in an expert
- Degrees and affiliations
- Be willing to work outside your area of expertise
- Be willing to compromise your "academic freedom"
- Deliverables
- Claim your independence
- How to discuss your relationship to industry
- If industry ties come to light
- A word on defection

Case: How the Food Industry Communicates on Obesity

4.
STRATEGIC COMMUNICATION

- Stay out front
- But do not hesitate to imitate
- Pro-science and policy positioning
- Victim positioning
- Reinvention
- Have third parties take the difficult stances
- Communication products
  - -Websites
  - -Search engine optimization
  - -Scientific journals
  - -Press releases
  - -Op-eds
  - -Op-ads
  - -Public letters, petitions, and pledges
  - -Book reviews
  - -Speaking engagements
  - -Museums

Case: Social Media "Addiction"?

5. CHALLENGE THE PROBLEM

- Hide or destroy internal evidence of the problem
- There is no problem
- Is there a problem? We are looking into it
- There is a problem, but it's a small problem
- There are bigger problems
- There is no longer a problem
- Change language to eliminate the problem
- Change statistics to eliminate the problem
- Change the scale of analysis to minimize or eliminate the problem
- There is a problem, but people are better off not knowing about it
- There is a problem, but it's not the Corporation's fault

Case: Questioning the Relationship Between Vaping and Covid-19

6. CHALLENGE CAUSATION

- Hide or destroy internal evidence of causation
- There is no evidence of causation
- Evidence of causation is weak, insufficient, or uncertain
- Appeal to "natural" properties
- Change statistics to eliminate causation
- Animal experiments and causation
- Change the scale of analysis to minimize or eliminate causation
- Something else causes the problem
- There are bigger causes of the problem
- Consumers are responsible for the problem
- Workers are responsible for the problem
- The government is responsible for failing to prevent the problem
- Our product causes a problem, but it is small and contained
- We cause the problem, but we are getting better
- The benefits outweigh the risks
- There are risks, as with everything

Case: Bias in the EAT-*Lancet* Commission

7. CHALLENGE THE MESSENGER

- Discourage internal dissent
- Offer an inducement
- Investigate individuals
- Legal intimidation
- Other forms of intimidation
- Claims of scientific misconduct
- Claims of bias
- Claims of alarmism
- Claims of being boring
- Claims of lying
- Claims of incompetence
- Claims of ulterior motives
- Ad hominem attacks

Case: Stalling the Preservation of Antibiotics for Medical Treatment Act

8. CHALLENGE THE POLICY

- More policy research is needed
- Sufficient policy is already in place
- The policy is too expensive
- The policy is a waste of taxpayer money
- The policy will hurt workers
- The policy undermines consumer freedom or individual rights
- The policy hurts poor people
- The policy costs lives
- The policy will be ineffective
- There are more important policies
- The policy is arbitrary
- The policy will have unintended consequences
- The policy threatens a collective identity or national sovereignty
- The policy is unnecessary because there is a technological fix
- The policy should not be determined by this group of people
- Any policy is overreach

Case: Human Nature Can Help Justify Inaction

9. OUTSIDE OPPORTUNITIES

- The problem is a result of human nature
- The problem is very complex
- The opportunity to address the problem has passed
- The next generation will address the problem
- Focus on household consumers
- Looking for villains is unproductive
- Criticism of universities and professors
- Social change happens slowly
- Technological solutions

Case: Growing Distrust in Big Business

10. NEAR-TERM THREATS

- Internal strife and talent retention
- Student activism opposing Industry funding to universities
- University disclosure policies
- Media policies
- Disinformation policies
- Lawsuits
- University programs and projects

APPENDIX

GLOSSARY OF TERMS

NOTES

EXECUTIVE SUMMARY

Every executive should own a copy of *The Playbook* and hope never to have to use it. But if there comes a time that scientific knowledge poses a risk to business operations, *The Playbook* is a guide on whom to hire, how to recruit experts, tips for effective communication, and ways to successfully challenge the science, the policy, and the scientists, reporters, and activists using science to further their policy agendas.

The Playbook is for every company. It outlines a universal strategy that is both offensive and defensive and is the predictable result of the central aim of the corporate structure—the pursuit of profit. *The Playbook* highlights achievements from a century of delivering strong financial performance in the face of challenging scientific "evidence" and how to modify scientific standards of evidence to outmaneuver attacks. It identifies useful arguments made by outsiders that companies can amplify. It also lays out near-term threats. Case studies related to upcoming material are provided before each section to help refine practical skills.

The business case for challenging scientific evidence that implicates a product in a social problem is straightforward. By delaying costly and intrusive science-based regulations, the creation of scientific disagreement buys time and saves money. As with many other cost-saving operations that are perfectly legal, but are nevertheless socially undesirable—such as cutting wages, moving manufacturing to countries with fewer regulations, using offshore tax havens—the unmaking of scientific agreement must be treated with discretion. Keep *The Playbook* confidential.

A successful campaign begins with a powerful arsenal. For tasks beyond the expertise or moral inclination of the internal workforce, there is outsourcing. A network of third-parties—lawyers, reporters, experts, public relations firms, think tanks, nonprofits, and trade associations—is necessary to mount a solid defense. *The Playbook* also includes a recruiting tool for attracting and cultivating university researchers who can challenge scientific research with a patina of independence. The assembled network will be equipped with a variety of tools, such as press releases and advertorials, and various arguments and rhetorical devices. This arsenal provides a defense against any aspect of the scientific process that implicates a product in a problem. The arsenal is deployed to carry out the four-pronged strategy: 1) challenge the problem, 2) challenge causation, 3) challenge the messenger, and 4) challenge the policy.

Disputing a problem identified using the scientific method can be done with varying levels of intensity. In some cases, the problem may be denied outright. In others, the problem may disappear under scientific reanalysis, it may be shown to affect only a small area or number of individuals, or the problem may arguably be so complicated that it is obvious that more research (and time) is needed.

If forced to accept the problem, the option remains to challenge the science of causation. Call into question the experimental design, the data, the methods, or the statistics. Emphasize alternative causes and fund studies that provide counterevidence. Focus on scientific uncertainty, the lack of scientific consensus, and the scientific dissent. If there is no genuine disagreement, create it. Know that any standard of scientific evidence can be disputed.

For scientists, activists, and reporters whose work will ultimately put business operations at risk, it may be necessary to call into question their reputation. Claim they are apocalyptic, biased, doom and gloom, hysterical, or radical. Intimidate or coerce them. These tactics have the added benefit of discouraging young professionals from asking similar questions (the so-called "chilling effect").

If the weight of scientific evidence is beginning to lead to policy discussions, challenge proposed regulations in much the same way as the problem or its cause. Claim that any regulation represents government overreach. Prolong the debate about which policy is most effective for as long as possible. Again, this buys time and saves money.

The overarching goals and strategies around challenging scientific knowledge that threatens business operations have remained the same, but some of the maneuvers, along with the media environment and culture, have been modernized. Many of the activities that companies previously carried out themselves are now subcontracted to public relations firms, law firms, and trade associations, similar to how manufacturing companies have sold off the parts of their supply chain with the greatest liability to middlemen. *The Playbook* will inspire thinking on how circumstances might change yet again.

There will also be opportunities to boost independent arguments that reinforce the position that government regulation is difficult, damaging, or futile. For instance, independent experts unrelated to any business or industry may claim that some problems, like climate change, are too complex to solve. Others may insist that policy inaction is the result of various failings by scientists, such as their lackluster communication skills. These outside ideas help buttress the mission to postpone regulatory action.

Finally, near-term threats are identified that could jeopardize the ambition of these efforts, ranging from internal conflict within a workforce, high-stakes litigation, government firewalls that prevent industry involvement in decision making, rumors of a new Museum of Agnotology devoted to educating the public about the creation of ignorance, and a new manuscript that reveals many of the details of *The Playbook* (see Appendix). However, there is every reason to believe free enterprise will continue to influence, control, and unmake knowledge generated through the scientific process for years, if not decades. *The Playbook* will help ensure that success.

THE PLAYBOOK

CHAPTER 1

DENIAL: A FIDUCIARY DUTY

The Corporation was created in the late sixteenth century through a series of government charters. As its potential to create vast amounts of prosperity and meet society's needs grew, the Corporation gained autonomy. A state law at the turn of the twentieth century allowed U.S. corporations to define the scope of their own charters without government oversight, which resulted in the variety of corporate forms that exist today, each with their specific advantages and all with the common goal of maximizing profit.

The modern Corporation is the backbone of the economy, with its employment, tax base, and unparalleled wealth creation for shareholders, executives, and charities. The Corporation is responsible for the economies of scale of assembly lines, the implementation of the Green Revolution, the efficiency gains from cheap energy, and the public health improvements from the innovation and manufacturing of lifesaving devices and medicines, including vaccines. Many believe there is no force more powerful to solve any crisis that humanity faces.[1]

The Corporation cultivates an image as "human, benevolent, and socially responsible"[2] although its core purpose is the pursuit of profit. Professors may debate[3] the extent of the Corporation's legal obligations to maximize profit, but there is no doubt that financial returns are the primary expectation of shareholders and executives. The Corporation can voice public support for other social values, but that support cannot come at the expense of the fundamental responsibility of financial performance.

As society's most powerful engine of prosperity, it will come as no surprise that the Corporation is always at risk and is always under threat. A study of more than 25,000 publicly traded companies in North America between 1950 and 2009 found the average company lifespan is just ten years.[4] Regularly discussed risks to the Corporation include market competition, globalization, unemployment, cyber attacks, and the unavailability of finance. But another risk can contribute to the Corporation's downfall: scientific knowledge.

SCIENTIFIC KNOWLEDGE

Modern scientific knowledge, which dates back to the seventeenth century, is a way of producing knowledge that uses a process of observation, hypothesis formulation and testing, and results. These methods and results should be reproducible by others. Scientific research is often written up and reviewed by several experts before it is published in a scientific journal. There is no single "science" and there is no "the science." Instead, there are many different views on how to approach questions in a scientific way.

Scientific knowledge is just one form of knowledge, but it is arguably the most powerful and most trusted. Science is "supreme among belief systems in its ability to create new knowledge."[5] Science is "certainly the most reliable body of natural knowledge we have got"[6] and "the most reliable deliverer of knowledge society has ever known."[7]

Among the many achievements of science is its ability to uncover the existence and causes of human-made problems. The scientific process can lead to discoveries that challenge intuition—for example, that a communal water pump can spread cholera; giving teenagers certain antidepressants can actually increase their risk of suicide; the insecticide DDT can weaken birds' eggshells; industrial refrigerants and solvents can deplete ozone in the atmosphere. Scientific knowledge can also precipitate government regulations, as it did in each of these cases.

Therein lies the risk of scientific knowledge.

Science can establish possible harm caused by the Corporation's product or its means of production, that can in turn catalyze a change in consumer preferences or, worse, government regulation. New rules inevitably increase the cost of production and reduce revenue or market expansion.

On the one hand, the reliability of science makes scientific knowledge potentially dangerous to the Corporation. On the other hand, a fundamental principle of scientific knowledge is that it is always open to revision. This revisionist quality is what makes science dependable over long time periods, but it also creates opportunities to challenge science in the short term. Science takes all pushback seriously, regardless of motive.

Fiduciary duty obligates the Corporation to dispute scientific knowledge that threatens operations. While the Corporation may not be able to control the scientific process or consensus forever, significant delays can be achieved through funding scientific experts to dominate an academic discipline or introduce scientific controversy.

Science can be used to challenge science, in part because science is never 100 percent sure of anything and always open to adjustment. Most features of scientific evidence can be easily and legitimately questioned. Which research questions are asked, how data are interpreted, the assumptions built into models, alternative hypotheses, uncertainty, confidence, the strengths and weaknesses of randomized controlled trials, the standards of statistical significance, possible confounding factors—almost every aspect of the scientific process presents an opportunity to refute independent research and a chance to show that the facts remain unsettled.

Scientific knowledge can be made and it can be unmade.[8] Progress gained toward scientific consensus can be lost. When scientific disagreements emerge, that division can undermine public confidence in experts or the science, and help buy time against burdensome regulations. *The Playbook* will help with the execution of this strategy.

DENIAL

Some professors have referred to attempts to challenges of scientific knowledge as "normatively inappropriate dissent"[9] while others use the simpler term "denial." Denial is most typically understood as a psychological reaction by an individual person, but, for the purposes here, that particular form of denial will be referred to as rejection.[10] Many instances of rejection have nothing to do with scientific knowledge—like a refusal to accept the existence of the Holocaust or that Barack Obama was born in the United States. Some scientific ideas that individuals reject—the safety of vaccines, or evolution by natural selection—did not originate with the Corporation.

Top-down, coordinated activities to discourage public acceptance of scientific knowledge—destroying, suppressing, concealing, challenging, or countering scientific evidence—are referred to here (in confidence) as denial. It is understood that some industries may consider these activities as "creating doubt" about a problem or a cause "without actually denying it."[11] The fact there is no broadly shared social understanding of the difference between doubt and denial or how to define denial is an advantage and there should be no attempt to clear up any confusion. However, in *The Playbook* the terms are clear: rejection is a psychological state; denial is a business operation.

Using these definitions, it is possible that the Corporation's employees do not personally reject a particular strain of scientific knowledge that they may nonetheless play a role in denying. Those hired to be part of a network to challenge scientific knowledge need only see themselves as doing their job or, in some cases, delivering value to their firm, clients, or shareholders. Individuals involved in the denial of scientific knowledge may come to see their participation simply as an obligation or a form of role-play, as opposed to genuine advocacy for an epistemic truth.

THE DENIAL OF SCIENTIFIC KNOWLEDGE

Denial of scientific knowledge that poses a risk to the Corporation may begin with preventing certain questions from even being asked. At times, "unsettling knowledge is thwarted from emerging in the first place, making it difficult to hold individuals legally liable for knowledge they can claim to have never possessed."[12] Pedants refer to this tactic as "strategic ignorance."

Other times, denial involves destroying and suppressing internally generated scientific knowledge that implicates the Corporation. The destruction or concealment of internal knowledge is easier than destroying or suppressing knowledge that was created outside the Corporation. Manufacturers of vinyl chloride conducted their own studies on animals that showed vinyl chloride caused cancer, which those manufacturers could (and did) prevent from being publicized. When independent research later showed potential harms associated with vinyl chloride, the manufacturers were not able to destroy outside results, but they could pose a challenge to them. The vinyl chloride manufacturers questioned the utility of animal studies, despite having conducted animal studies themselves, and insisted that long-term epidemiological studies on humans be conducted before regulations be considered. They suppressed internal knowledge, challenged external knowledge, and successfully postponed burdensome regulations.[13]

A large portion of scientific denial may be characterized as bullshit. As philosopher Harry Frankfurt pointed out, "the essence of bullshit is not that it is *false* but that it is *phony*." Bullshit is "produced without concern with the truth" and therefore may or may not be true.[14]

In the early 1950s, several research studies in influential medical journals called attention to the link between cigarettes and lung cancer. Due to "the grave nature of a number of recently highly publicized research reports on the effect of cigarette smoking"[15] and the fact that cigarette sales showed a decline for the first time in more than two decades,[16] cigarette manufacturers met in December 1953 to mount a collective defense against the attack from medical knowledge. They decided to hire public relations firm Hill & Knowlton to defend the industry. On their side, Hill & Knowlton did not independently assess any medical research or seek opinions from experts.[17] Hill & Knowlton was there to support the goal of their clients, the cigarette companies, in a way that aspired to be "conservative and long-range" without "any flashy or spectacular ballyhoo."[18]

The cigarette companies funded the denial of scientific knowledge, and hired a public relations firm to administer the bullshit. Hill & Knowlton challenged the science using tactics that were real but nevertheless phony, because the central objective was not scientific truth, but to create scientific confusion and regulatory delay on behalf of its clients. What was scientifically true was only relevant insofar as it served the objective to restore the sales of cigarettes.

DENIAL PAYS

Denial is an investment. Delay is the deliverable. The measure of success is the degree and duration of government gridlock—the "holding strategy" or the "not so fast" strategy[19]—that permits continued sales of the product.

Regulations can be so costly, leading to increases in production costs, forgone revenues, or both. After scientific research implicated neonicotinoids (a type of insecticide used on crops like soybeans) in the decline of bees, the European Union restricted their use in 2013. According to the CEO of Switzerland-based pesticide manufacturer Syngenta (now owned by Bayer) at that time, the new regulations led to a loss of $75 million for the company that year.[20] (That Syngenta did $13.6 billion in total sales in 2019 is beside the point.)[21] An executive for Luzenac America, the U.S. branch of a French talc mining company, said that if the National Toxicology Program's Report on Carcinogens listed talc in their report there would be "devastating consequences for the talc market worldwide." For Luzenac, the company would "see a virtual immediate loss of our sales to the personal care market—around $10 million in sales the first year." Luzenac would also "likely suffer a deterioration of sales in all markets" in subsequent years and "civil litigation would likely skyrocket."[22] When a science-based regulation is likely to impose tens of millions of dollars in costs and snowball into more regulations and lawsuits, the justification for allocating funds to deny the science is clear.

Denial of scientific findings makes economic sense. The denial of climate change, arguably the boldest scientific denial that exists, cost at least $9.77 billion[23] from 2003 to 2018, but the payoffs, including effectively zero legally binding international policy, legitimize the expense. Cigarette manufacturers were under threat by science for decades before they lost their first court case in the 1990s. Industry-funded studies indicating that the chemical bisphenol A (BPA) was not likely to cause harm (despite independent research that showed otherwise) bought regulatory delays in the U.S. and the European Union.[24] Although denial efforts can be expensive, the costs can be shared by an entire sector or even several sectors.

There are three preconditions for scientific denial: 1) a company or industry for which 2) scientific knowledge 3) poses a genuine regulatory threat. If there is scientific knowledge, but no threat of science-based policy, there is little need for companies to spend money challenging science, just as if there are no taxes, elaborate tax avoidance efforts are unnecessary. The oil and gas industry has not worked to deny climate change outright in Brazil, because the oil and gas sector is not as economically powerful there as it is elsewhere. Instead of denying the link between fossil fuels and climate change, denial occurs about the role of the beef industry, which makes up almost 10 percent of the Brazilian economy.[25]

Agrochemical companies are actively leading efforts to challenge scientific authority in Argentina, the world's third-largest exporter of soybeans. Scientific knowledge and its denial are ultimately geopolitical.

A SHORT HISTORY OF SCIENTIFIC DENIAL

Scientific denial became an important part of business operations beginning in the twentieth century. Prior to that point, it was up to the manufacturers to know whether their product was harmful, which primarily meant harmful to workers. If the Corporation did not document an increase in disease from job-related activities in its workforce, regulations were unlikely.[26] Denial was unnecessary because the only oversight of the Corporation's affairs came from the Corporation itself. But the twentieth century saw an increase in government involvement.

The first wave of corporate-led scientific denial occurred in the early twentieth century over worker health and safety due to handling of hazardous materials, including radium, lead, and asbestos. Physicians were the main source of expert scientific understanding. By the mid-1920s, seven U.S. Radium Corporation employees had died from using the radioactive glow-in-the-dark paint, but U.S. Radium denied that the cause of their deteriorating bones was radium. In the 1920s, some refinery workers who had been exposed to lead as a new additive to gasoline showed concerning symptoms (such as memory loss and convulsions) and a handful eventually died. The Lead Industries Association endeavored to "correct misstatements" and "calm misapprehension" through the 1930s,[27] and companies continued selling leaded gasoline and paint, which lingers in households to this day.

Then the asbestos industry elevated the efforts to deny scientific knowledge. Several asbestos companies discovered in the 1920s and 1930s, through their own medical doctors, that workers who repeatedly breathed in dust from asbestos, a mineral fiber used as insulation, developed scar tissue in their lungs and some workers eventually suffocated. It was standard practice that company doctors would not tell a worker about signs of asbestosis (an illness for which the cause was regrettably made explicit in its name). Industry lawyers deleted passages from company medical documents that linked asbestos work to sickness. Johns Manville, the largest North American asbestos producer, suppressed a 1943 report that confirmed the link between asbestos and cancer. Compensation claims filed during that time period were settled out of court with a secrecy order and, due to both the social context of the Depression and the industry suppression of information (which insurance companies abetted), the first attempts at asbestos litigation were unsuccessful and then forgotten.[28] Then, decades later, in 1982, more than 16,000 workers threatened asbestos-related claims. Johns Manville was forced to declare bankruptcy but the lesson remained clear: denial bought the asbestos industry nearly five decades of regulation-free profits.

A second wave of denial occurred over consumer products, especially tobacco and pharmaceuticals. Cigarette manufacturers conceived of scientific denial as a "holding strategy"[29] and "sand in the gears."[30] They achieved a half century without regulation. The manufacturers even accomplished a short-term increase in consumption in the U.S., where smoking grew 20 percent between 1954, when tobacco companies launched their denial efforts, and 1961 (387 billion cigarettes a year to more than half a trillion).[31]

A third wave of denial occurred in response to the environmental and animal protection movements, and can be seen in the energy sector, the chemical manufacturers, forestry, and food production, including the meat, dairy, and aquaculture industries. Fossil fuel companies and their network of allies benefited from tobacco's experience and likewise achieved regulatory delays, all while forging their own new methods for denying scientific knowledge. Global greenhouse gas emissions were the highest in world history in 2018—thirty years after the first testimony on greenhouse gases and climate change before the U.S. Congress (and thirty years after the establishment of the Intergovernmental Panel on Climate Change). Just like cigarette manufacturers, fossil fuel companies put aside their differences, their belief in competition, and their employees' personal moral standards, and came together to fight the threat of scientific knowledge and its impact on public perception and, ultimately, government policy.

A RANGE OF DENIAL

Sometimes, the stakes are high enough to justify extreme denial. A prime example is the bold campaign to deny the existence of climate change, which threatened one of the largest and richest industries in the world. Energy producers came together to challenge not only the cause of climate change, not only the people studying it, and not only the policies, but the entire problem of climate change itself. The denial of climate change required titanic investment and coordination but offered handsome payoffs.

Other times, denial may be more moderate. Merck Pharmaceuticals sponsored research that misrepresented the safety of its arthritis relief medication Vioxx, which caused heart attacks and strokes in many of its users. It allegedly paid scientists and its sales representatives to also misrepresent the drug's safety.[32] Once medical research revealed that Vioxx was unsafe, Merck also allegedly tried to suppress research on the dangers of Vioxx. However, Merck never went quite so far as to launch a campaign to deny the publicly available scientific findings (they did take Vioxx off the market in 2004 and later pled guilty to criminal charges, but they expressly denied that they misrepresented the drug's safety or suppressed research).[33] There are degrees of denial, and the Corporation must choose the best fit for its situation, while always adapting to cultural shifts and new political and technological landscapes.

FOCUS ON THE SHORT RUN

It is important to stay focused on the short-run benefits of denial. There are long-term risks associated with denying scientific knowledge. The Corporation might jeopardize institutions, including trust in science, the media, and possibly even democracy as a political system. Scientific denial may result in compromises to human health. Many industries, including tobacco,[34] have worked to undermine the public trust in national health organizations, which may mean less preparation for a global health crisis.

Philosopher of science Thomas Kuhn noted that nothing guarantees that science will go on forever. Kuhn said in a 1991 interview, "There was a beginning to [science]. There are lots of societies that don't have it. It takes very special conditions to support it. Those social conditions are now getting harder to find. Of course it could end."[35]

Would losing the authority of science really be that bad? Fortunately, the question about the large-scale, social effects of undermining scientific knowledge are not central to the Corporation's immediate financial performance. Corporate executives are under immense pressure to deliver quarterly earnings, not to save science, democracy, or the planet.

Long-term implications of denial campaigns lie outside the Corporation's singular and short-term goal to generate revenue, and they should lie outside any individual employee's purview, too. The Corporation must convey, preferably implicitly, that scientific denial is part of fiduciary duty, and that duty takes precedence over any individual employee's morality. Denial can be easily rationalized by executives, as well as by most employees. A former manager for Johns Manville, the asbestos manufacturer, explained, "At Manville, denial became endemic to the corporate culture, so much so that even after top executives had recognized health and safety as a critical issue, many middle- and lower-level managers continued to hide behind rationalizations."[36]

Research has shown that the Corporation can create a culture and incentive structure that is maximally conducive to scientific denial and overrule to a large extent any individual's belief in what is right or wrong. The banking sector has successfully created a culture that encourages dishonesty for monetary gain. An experiment involving hundreds of Europeans showed that bankers are more morally conflicted than other social groups. If they did not think about their professional identity, bankers were just as honest in the experiment as the general population, but when their "banker" identity was activated, bankers were less honest than prison inmates.[37] The culture of the workplace can hold remarkable sway over an individual employee's conscience.

DISCRETION

The denial of scientific knowledge is a delicate subject that must be treated with discretion. It will never be addressed in a TED talk, at a roundtable at Davos, or in the pages of *Fortune* magazine. *The Playbook* is strictly confidential for use only by trusted members of the Corporation.

Any evidence of executing the tactics outlined here should be contained and eventually destroyed. The discovery by the public that there has been an intentional effort by the Corporation to undermine or suppress scientific knowledge can lead to backlash. Judges and juries who may eventually be called upon to assess punitive damages may be angered by evidence that the Corporation in fact had internal knowledge of potential harms, yet denied outside knowledge of those same harms. Energy companies like Exxon, BP, and Shell are under attack from claims that they mounted a long-term effort to oppose action to mitigate carbon emissions while having internal knowledge of climate science.[38] Car manufacturers like General Motors and Ford are in a similar predicament.[39] DuPont settled thousands of lawsuits alleging that one of their chemicals (perfluorooctanoic acid) had contaminated Ohio and West Virginia drinking water.[40] Part of the case rests on the claim that DuPont knew the chemical caused disease and concealed that knowledge.[41]

Deny, and then destroy or conceal evidence of denial—a form of second-order denial. As one historian put it, first falsify the science, then falsify the history.[42] If the topic ever comes up, follow the lead of the General Motors spokesperson who responded to allegations in 2020 about the company's previous denial of scientific knowledge with: "There is nothing we can say about events that happened one or two generations ago since they are irrelevant to the company's positions and strategy today."[43]

There have been unfortunate moments when denial-related documents have been published through whistleblowers, leaks, or during pretrial discovery and legal settlements. Four thousand pages of historical documents from Brown & Williamson Tobacco Corporation made their way to a university professor, who gave them to *The New York Times*.[44] The disclosure of 40 million pages of tobacco industry documents through the Master Settlement Agreement was also a setback for cigarette manufacturers. One early climate change denial campaign was dissolved after memos related to the public relations campaign were leaked to the press.[45]

Those who work to expose the denial of scientific knowledge, including activists, journalists, historians of science, and some sociologists, have been cause for concern. The activists running the Climate Investigations Center have beleaguered the fossil fuel industry. Monsanto-obsessed reporter Carey Gillam has tormented chemical manufacturers. Historians of science Erik Conway and Naomi Oreskes harassed the handful of Cold War scientists who denied science on behalf of many industries with their book on the subject. Robert Proctor has viciously attacked the cigarette manufacturers, has named the university historians who had testified on behalf of the tobacco industry,[46] and in his book *Golden Holocaust* drolly thanked "those many cigarette industry lawyers [fourteen in total] with whom I have sparred over the past dozen-odd years."[47] Historian Stanton Glantz glibly claimed that his years exposing the tobacco industry's tactics prevented a midlife crisis.[48]

Fortunately, the number of activists, reporters, and professors interested in this subject is few and there is no institutional support to incentivize researchers who work on denial to act together in any strategic way. To date, there is no top-ranked university that has created a department where experts convene to study scientific denial or consider how to protect science from the interests of free enterprise. Only one major news organization has a designated "disinformation" unit. There is no undergraduate program of study that prepares students for denial and disinformation. There is a course here and there (such as Carl Bergstrom and Jevin West's "Calling Bullshit"), a few books, a few dozen scientific articles, and some long-form reporting. But the work has not coalesced into the power of an actual field of research or public policy. Nobody has gotten out in front of the Corporation. Everybody is reacting. This is all for the best (although see the chapter on Near-Term Threats).

As a result, citizens are naive and often fail to discern or prevent the tactics of denial. Students at Princeton University who heard fossil fuel industry ally[49] and professor emeritus William Happer speak in 2017 told their professor afterward that they did not know how to respond to scientists like him who question climate change.[50] This kind of paralysis among students at an elite university is testament to *The Playbook*'s achievements. The more people understand about denial strategies, the harder it will be for the Corporation to achieve regulatory gridlock.

On a somber note, sometimes a denial effort fails. The trade association for the nuclear energy industry hired the best public relations firm and a polling and marketing research firm to launch an expensive campaign in 2006 in an attempt to win over public support for the Yucca Mountain Project, a repository for nuclear waste. Opposition to the project actually grew and plans for the repository were abandoned in 2011.[51] Climate change denial has not yet taken strong root in Germany, although many have tried.[52] These failures present learning opportunities.

There are also rare cases when employees or even entire companies rebel. In many cases, subversion can be quietly and effectively addressed. After an Exxon employee pointed out in a 2020 meeting with managers and executives the contradiction that Exxon had publicly acknowledged the need to reduce emissions, but nevertheless had internal plans to grow emissions, he claimed he received a poor performance review and says he was pushed out of the company.[53]

Other times, an insurrection can be disruptive. In 1997, BP broke ranks with the other major Western energy companies and admitted to the existence of man-made climate change. In 2015, a handful of top executives left Edelman, the world's largest public relations firm, because of Edelman's representation of fossil fuel companies and climate change denial, which created a public relations issue for the public relations firm (Edelman caved in to the pressure and says it no longer represents fossil fuel companies).[54] In the mid-twentieth century, Dr. Kenneth Smith, a former doctor for the asbestos manufacturer Johns Manville, went against company advice and gave a deposition stating that the company indeed knew of the hazards workers faced because workers received regular X-rays. The lawyer who deposed the doctor said, "As for Dr. Smith's motive in revealing all this incriminating evidence, I can only tell you what I surmise. I think that his conscience was bothering him and that he wanted to set the record straight."[55]

Fortunately, the Corporation is not bound by a conscience. The denial of scientific knowledge is simply part of fiduciary duty. An aspect or even an entire campaign to deny scientific knowledge may sometimes fail. More often, though, the Corporation's relentlessness, a reflection of its fiscal supremacy, will lead to emphatic victory against the risks of scientific knowledge and delay any related regulations.

CASE: DIVERSE ALLIES IN THE ENERGY SECTOR

Gender and racial equity are socially desirable, and the energy sector has promoted collaborations that are diverse and inclusive. Shell Oil hired public relations firm Edelman, which in 2018 hired the actress Letitia Wright, star of the movie *Black Panther*, to highlight some efforts by young women engineers.[1] The natural gas sector created the website Women for Natural Gas, which featured a chronology of women in the petroleum industry under a tab labeled "Herstory" (the website was later criticized for using stock photos to represent the women who had contributed testimonials; one of the headshots was taken from a senior employee at LinkedIn).[2]

Western States and Tribal Nations Natural Gas Initiative is a nonprofit organization that works to develop natural gas markets, pipelines, and terminals. Its website declares the group is "led by sovereign tribal nations," and emphasizes its mission is to facilitate "tribal sovereignty" and "promote tribal self-determination," even though of the twenty-plus members, only the Ute and Southern Ute are tribes, and most of the other members are industry groups.[3] The organization's president, as listed on its 2019 tax filing,[4] is also president of Consumer Energy Alliance, a trade association for the energy industry.

SoCalGas, the largest natural gas distribution utility in the U.S., makes charitable contributions each year, including large donations ($107,750 between 2014 and 2018) to Pacoima Beautiful, a grassroots environmental justice group that works in the mostly Latino northeast San Fernando Valley. SoCalGas has also donated money to the American Indian Chamber of Commerce of California ($10,500), the California Latino Leadership Institute ($10,000), and the Greater Los Angeles African American Chamber of Commerce ($28,000), all of which then provided letters of support[5] for a SoCalGas policy proposal that would give customers the option to purchase natural gas from emissions from the agricultural and waste sectors.[6]

What allegiances could be built with groups working on behalf of popular social causes?

CHAPTER 2

THE ARSENAL

To prepare to challenge scientific knowledge that could lead to public disapproval, litigation, and, especially, government regulation, the Corporation must build and mobilize an extensive arsenal made up of individuals, institutions, and communications networks. In most cases, this extensive arsenal of people and firms will ascend to the status of an integrated "web" or "network." The members of this network are encouraged to conceal or downplay any ties to the Corporation. As an added benefit, the organizations hired to promote and defend the Corporation are likely to refuse to accept other outside contracts that could compromise or challenge the Corporation's accounts (for instance, a public relations firm with a coal account will turn down an environmental group with a proposed campaign to attack coal).

Although the Corporation may compete with other companies in the marketplace, it can be cost effective when an entire sector comes together to address scientific knowledge that threatens the industry. Cigarette manufacturers shared "the expense of the public relations program" that launched their denial campaign, which was "prorated in accordance with the relative volumes of business of the several companies."[1]

The web or network, which includes law firms, think tanks, public relations firms, political consultants, and pro-business media outlets, will create a constellation of counterpoints to scientific knowledge. Each actor will use a unique approach and leave a different trail of evidence that will make it difficult to reconstruct the larger whole. In the latter half of the twentieth century, the cigarette manufacturers provided $450 million to the Council for Tobacco Research, which led to more than 7,000 scientific papers, many of which challenged the relationship between smoking and cancer.[2] To defend the herbicide glyphosate, the agrochemical industry created a network of professors, front groups, and think tanks who appeared to be independent.[3] The fossil fuel industry's network to challenge climate change included at least 4,556 individuals with ties to 164 different organizations.[4]

EXECUTIVES

The Corporation requires leadership with grit and good instincts. Executives in the cigarette business are prime examples. Edward A. Darr, president of R.J. Reynolds, initiated the "no real proof" approach to the growing body of knowledge linking cigarettes to health problems and he called doctors who linked cancer to smoking publicity hounds.[5] Paul M. Hahn, president of the American Tobacco Company, became "concerned with the highly publicized claims of certain medical men" and sent the telegram on December 10, 1953, that initiated the first meeting at the Plaza Hotel in New York.[6] Eight tobacco executives plotted how to defend their industry against the scientific evidence linking cigarettes and cancer, the first domino to fall in one of the most sophisticated denial campaigns. Hahn's successor, Robert B. Walker, told *The New Yorker* a decade later: "If we were convinced that cigarettes were harmful, we wouldn't be in the cigarette business."[7]

Fossil fuel industry executives have likewise shown leadership. Both Lee Raymond, who led ExxonMobil from 1999 to 2005, and his predecessor, Lawrence Rawl, broke ranks with Exxon's prior approach in the 1960s and 1970s of trying to work with the rising values of environmentalism. Preceding Rawl, Exxon's CEO Clifton Garvin had tried to bend to environmental values and had even installed solar panels to heat his swimming pool at his house in New Jersey. In contrast, under Raymond's clear-eyed leadership, ExxonMobil became more aligned with climate change denial, as well as with the Republican Party, and to a greater extent than any other large oil and industrial corporations. Raymond not only rejected that global warming was caused by human activity, but he also rejected global warming full stop.[8] Rex Tillerson, who took over as CEO of ExxonMobil after Raymond, shifted the company's position to accept the science of human-induced climate change, although he maintained climate change was not dangerous and continued to support broader industry efforts to deny climate change. Since 2017, Darren Woods has led ExxonMobil and he has claimed that the company supports zero-emissions targets[9] (while the company also continues to support climate denial).[10]

Executives should strategically join the boards of various institutions, including universities, museums, and nonprofit organizations, which may be useful at some point in shaping public opinion or preventing certain kinds of research. An oil executive and board member at Caltech once called on the university's president to discourage a Caltech professor from his research on industrial lead.[11] Before it was brought to the public's attention, David Koch of Koch Industries was on the board of both the Smithsonian and the American Museum of Natural History, where some believe he wielded influence over museum content. A 2015 exhibit that touched on climate change in the Smithsonian's Hall of Human Origins, which received a $15 million donation from Koch, did not include any reference to human-caused greenhouse gas emissions.[12]

TRADE ASSOCIATIONS

A trade association brings together companies with a common interest and can act as a hub of coordination. They are also ideal instruments for challenging science and policy. Trade associations can do work on behalf of an entire industry so that no single brand appears to be responsible. Any message that the Corporation prefers not to communicate directly, a trade association can deliver on its behalf. There are thousands of trade associations, some with straightforward names, such as the National Mining Association or the Lead Industries Association. Some names suggest activities that are beyond the scope of the trade association, like the National Fisheries Institute, which might present like a government agency or a research center, but is in fact the trade association for the U.S. seafood industry.

The Corporation may also want to join a large national association, such as the National Association of Manufacturers or the U.S. Chamber of Commerce, that can work to block regulations. Coalitions can also be created ad hoc when there is a common cause and something to be gained from cross-sector collaboration. In 1989, the American Petroleum Institute (a trade association) created the Global Climate Coalition (a coalition) to collaborate with aluminum and autoworkers on a specific campaign to block climate policy.[13]

PUBLIC RELATIONS

The Corporation usually has its own internal public relations department. As early as 1886, AT&T hired its own publicist to work on the company's image and reputation.[14] In the early twentieth century, external public relations firms were created to manage crises, influence public opinion, and lobby government officials on behalf of the private sector. The founders of public relations were focused on crisis management and on advertising. In 1914, John D. Rockefeller Jr. retained Ivy Lee to help improve the family's image after the violent repression of a coal miner strike in Colorado. In 1929, the president of American Tobacco hired Edward Bernays, a nephew of Sigmund Freud, to help increase cigarette consumption. Bernays, like his uncle, fully appreciated the role of the subconscious in manipulating public understanding and he participated in several notable marketing campaigns, including one that convinced women to smoke by equating cigarettes with sexual power.[15]

PR firms "exist to be advocates for [their] clients."[16] For any product, "the goal of the public relations campaign is to effectively communicate the history and safety . . . and its economic importance."[17] PR firms "address fears and fight misconceptions."[18] They can set up research groups that appear to be independent and challenge scientific consensus. They can increase the visibility of industry-funded experts.

Effective communication need not be honest communication. Unlike the field of journalism, which has historically adhered to a set standard of ethics about the kinds of information that are acceptable, PR firms have no such standards. They can create and communicate any information with little attribution or indication of its origins. PR employees are sworn to secrecy, and are trained in how to persuade as well as inform.

PR firms have honed their skills working across many different client portfolios. John W. Hill, founder of the eponymous Hill & Knowlton in 1927 in Cleveland, Ohio, was one of the PR executives most adept at challenging science. The Manufacturing Chemists' Association hired Hill's firm in 1951 "to combat the negative publicity the industry was receiving in connection to congressional hearings on chemicals in food"[19] and Hill successfully prevented the government from mandating testing of the food supply for chemicals.[20] Hill went on to become a mastermind in using the notion of scientific controversy to his clients' benefit.[21]

After cigarette executives met to discuss the threat of medical research in 1953, they agreed that "the feeling of those present [was] that the industry could most effectively face this problem by jointly engaging a public relations counsel." The work ahead was "of such a specialized nature that an advertising company could not deal with it with the delicacy that is required." Hill & Knowlton "was considered to have the necessary qualifications of high caliber and integrity and the experience to handle the assignment."[22]

It was Hill's idea for the tobacco industry to sponsor medical research—just not the kind of research that would investigate or, worse, confirm the relationship between lung cancer and smoking. According to Hill, "the tobacco account also gave Hill and Knowlton, Inc. expertise and personnel for dealing with scientific and medical problems in far better fashion than [they] had been previously able to do." It prepared them for "similar problems of other clients—notably the Licensed Beverage Industries, the Gillette Company, Procter and Gamble, and the Pharmaceutical Manufacturers Association."[23] In the late 1960s, Hill & Knowlton got the Johns Manville asbestos account. In the mid-1970s, the plastics industry, faced with new government standards with OSHA, the Occupational Safety and Health Administration, calling for "no detectable limit" on vinyl chloride, also hired Hill & Knowlton "to refocus public and congressional attention and to reshape the national debate about the effect of plastics on American society."[24]

PR firms have been essential to scaling and disseminating denial campaigns locally, nationally, and globally. The chemical manufacturer Syngenta hired the Chicago-based PR firm Jayne Thompson & Associates to counteract a 2004 lawsuit in Illinois aimed at making Syngenta pay to remove atrazine from drinking water. The PR firm recommended planting critical stories about the litigious nature of local courts.[25] Hill & Knowlton coordinated with the tobacco industry in Britain.[26] There are hard-to-trace PR firms that create fake accounts on social media to spread information that suits the Corporation's needs.[27]

CONSUMER ADVOCACY GROUPS

A new weapon was added to the arsenal in the 1980s: groups that appear to be grassroots advocates for consumers but are in fact connected to the Corporation, often created by PR firms.[28] Examples include the Center for Consumer Freedom (CCF), Consumer Advocates for Smoke-free Alternatives Association, and Americans Against Food Taxes. Some PR firms even have "an independent grassroots lobbying unit." In the 1990s, Monsanto hired Edelman to create a "grassroots group" that would oppose labeling foods that were genetically modified.[29] Syngenta also kept "third-party allies" on retainer who could appear to be independent supporters.[30] This hired network of people can be used to sign petitions, protest, and ask probing questions of politicians.

Consumer advocacy groups can also be used to sway representatives. A Virginia congressman and known swing voter received six letters from two advocacy organizations that claimed to represent minority communities opposing an upcoming vote in 2009 to advance cap-and-trade policy to address fossil fuel use.[31] The letters probably would have worked except that they were exposed as forgeries and linked to the PR firm Bonner & Associates, and then eventually to the coal industry. (Jack Bonner, then president of Bonner & Associates, astutely blamed a temporary employee for sending the faxes.)[32]

LAW FIRMS

A law firm can help the Corporation avoid complicity in denial inside and outside the courtroom. Law firms have "provided intimate counsel to the industry, directing research, cultivating experts, shredding or sequestering documents, and aiding and abetting in other ways."[33] Law firms have administered research funds to expert scientists and professors.[34]

COMMUNITY FOUNDATIONS

If law firms are not adequate for distributing financial resources, there are other ways to funnel money through intermediaries that can obscure the funding's origins and can help insulate the Corporation from overt associations with denial campaigns. In the mid-1930s, several silica companies set up the Air Hygiene Foundation to conduct research on silicosis (lung fibrosis caused by breathing in silica) in a "confidential manner" so that "no one would know what industries or individuals were contributing to the fund."[35] Donors Trust was established in 1999 as a public charity to "ensure the intent of donors dedicated to the ideals of limited government, personal responsibility and free enterprise."[36] After the likes of ExxonMobil and Koch Industries received bad press in the mid-2000s for directly funding groups that questioned climate science and opposed climate policy, companies switched to channeling funds to their networks through Donors Trust.[37]

PRIVATE INVESTIGATORS

Private investigators can be useful when it comes to challenging the credibility of scientists, reporters, and activists who deploy scientific knowledge for their purposes. Private investigators can acquire documents, intercept call records, trespass, and conduct surveillance. During a lawsuit related to the herbicide atrazine, its manufacturer, Syngenta, hired a detective agency to look into scientists on a federal advisory panel to the U.S. Environmental Protection Agency (EPA).

Keep such activities discreet. "I don't think it would be helpful if it were generally known that we research [government science advisers]," noted the head of Syngenta's communications in a company memo.[38] The Koch brothers, owners of the private industry conglomerate Koch Industries, hired a former New York police commissioner to look into Jane Mayer after her 2016 book *Dark*

Money (the only nugget to emerge from Mayer's lackluster past was that her great-great-grandfather worked at Lehman Brothers, which had done business with Nazi Germany).[39] When approached by Mayer, the lead investigator said, "I don't comment. I don't confirm or deny it." A Koch brothers spokesman also did not reply to questioning.[40] Silence on such matters is the best response.

EXPERTS

Individual experts can challenge scientific findings that jeopardize the Corporation's product or production method. Experts make public appearances, provide courtroom or congressional testimony, and produce white papers, social media, and peer-reviewed research articles. University experts are highly trusted by the public. A lobbyist for Monsanto who studied television ads found that third-party scientists were even better for convincing the public of the safety of genetically modified organisms (GMOs) than farmers who were also mothers.[41]

Research conducted by experts with industry support is manifold times more likely to produce favorable scientific results than research conducted by independent experts, which is why funding scientific research is a winning tactic. Time and time again, industry has benefited from producing its own scientific studies, with clear gains for drug and medical device manufacturers,[42] nutrition research,[43] sugary beverage manufacturers,[44] and indoor tanning salons.[45]

Occasionally, the Corporation will have the great fortune of finding individuals with ideological commitments that align perfectly with the Corporation's interests.[46] But, in general, there are very few outside experts who are independently devoted to the Corporation's cause. Test the waters by offering small contracts, consultancies, grants, and donations. If the expert's performance meets expectations, more funding can be provided. If the performance does not deliver, move on to better investments. There are three main venues from which to recruit expertise: think tanks, consulting firms, and the holy grail of expertise, universities.

Think tanks

Think tanks are perceived as scholarly institutes, but teaching and learning are not part of their mandate, which means think tanks are not beholden or sensitive to student politics or values. Think tanks were first incubated by the government—specifically the U.S. government during World War II. Their aim was genuinely to think, more specifically, "to think imaginatively" and envision "new ways of solving problems that would contribute to the war effort."[47] After the war, the U.S. government continued to fund the RAND Corporation in California, a think tank charged with imagining "new futures and new concepts of how society could work."[48] RAND led, for example, a fifteen-year research project that supported a reorganization of private health insurance.[49]

Think tanks became an arm of the private sector in the mid-twentieth century, championed by the economist Friedrich Hayek. Hayek's vision was a private "scholarly institute" to influence public opinion and advocate for free market ideology and he told many of his peers about the possibilities think tanks could provide, including Antony Fisher. Having made a modest fortune from England's first broiler chicken operation, Fisher used part of his wealth to set up one of England's first think tanks in 1955, the Institute for Economic Affairs (factory farming and think tanks both being natural outcomes of capitalism). Fisher ultimately established 150 or so other think tanks around the world to spread free market ideas.[50]

Think tanks try to capitalize on their earlier reputation as places of imagination, but many think tanks today ultimately exist to challenge regulations, draft legislation, fund advertisements, provide expert testimony, influence public opinion, and provide institutional homes for experts whose paychecks ultimately depend on the Corporation. After experts were criticized for receiving funding from the fossil fuel industry, many of them moved over to think tanks so that they no longer had direct industry ties, although the same fossil fuel companies funded the think tanks.[51] The Competitive Enterprise Institute, a libertarian think tank established in 1984, was open in the 1990s about its aim "to give intellectual aid and comfort to greenhouse skeptics in the U.S. and other delegations."[52] Think tanks sponsored more than 90 percent of the 141 English-language books that were skeptical of environmentalism between 1972 and 2005.[53]

Think tanks may have anodyne names, like the Marshall Institute and the Heritage Foundation. The Heartland Institute, funded in large part by ExxonMobil, Charles and David Koch, and later Donors Trust, has been called "the primary American organization pushing climate change skepticism"[54]—and also the primary organization pushing the boundaries. In 2012, Heartland sponsored a series of billboards that compared those who believe in global warming to "murderers and madmen," including the Unabomber, Charles Manson, and Osama bin Laden.[55]

Consulting firms

Consulting firms are where public relations and scientific expertise meet.[56] They produce the scientific results to fit the Corporation's needs, including "product defense" or "litigation support."[57] The "multi-disciplinary engineering and scientific consulting firm" Exponent, formerly named Failure Analysis Associates, was founded in 1967 by a handful of professors and engineers from Stanford University.[58] Exponent "has been best known for analyzing accidents and failures to determine their causes, but in recent years it has become more active in assisting clients with human health, environmental, engineering and regulatory issues associated with new products or processes to help prevent problems in the future." Exponent has expertise in "more than 90 technical disciplines" and notes that "over 50% of [its] staff hold a Ph.D. or M.D. in their chosen field of study."[59] When BP needed an expert to question research that suggested that the Deepwater Horizon oil spill had damaged corals in the Gulf of Mexico, whom did it call? An Exponent scientist and another consultant published a response in the journal *Proceedings of the National Academy of Sciences* arguing that there were "other plausible explanations, including the presence of natural oil and gas seeps near the coral."[60] When meat producers wanted to challenge the classification of processed meat as carcinogenic by an international agency, whom did they hire?[61] Exponent scientists published a number of studies between 2009 and 2014 that found no real proof of a causal link between processed meats and cancer. The National Pork Board hired Exponent experts to conduct a study that was published in *Nutrition Research* titled "Fresh and Fresh Lean Pork Are Substantial

Sources of Key Nutrients When These Products Are Consumed by Adults in the United States."[62] Consultants always deliver. However, because they are explicitly work-for-hire, consultants have lower status in terms of public trust in their expertise.

University experts

Finally, no arsenal is complete without some independent researchers with an affiliation or, better yet, a full-time position with a highly ranked university. These individuals are the ideal choice to lend eminence to the Corporation's message. So important are university experts that the next section lays out how the Corporation can recruit them and get the most out of the relationship.

CASE: EXPERT OPPOSITION TO DIVESTMENT

The campaign to divest from fossil fuels began at Swarthmore College in 2010 and is now a highly visible, organized response to climate change. Students around the world are trying to persuade their schools to divest their endowments, with charades like a 116-day sit-in in 2016 at MIT (where the full academic year is just 126 days). As of 2020, half of all the U.K.'s public universities had committed to divestment.[1]

Who better to make the case against divestment to overzealous university students than their own professors and peers? University professors such as Harvard's Robert Stavins and Princeton's William Bowen have published articles arguing against divestment in high-profile newspapers. After students persuaded Stanford University to divest from coal in May 2014, UCLA professor Ivo Welch wrote in *The New York Times,* "Tonight, the 20,000 students at Stanford will sleep more soundly. The world will be a better place. Except that it won't be." Based on his research related to divestment from South Africa during apartheid, Welch argued that "individual divestments . . . have never succeeded in getting companies or countries to change."[2] Paul Tice, an adjunct professor of finance at NYU's Stern School of Business, argued against divestment in 2018 in *The Wall Street Journal* because oil and gas are "too large, diverse and volatile a sector for major institutional investors not to own."[3] (In 2019, Tice wrote another piece for the *Journal* titled "On Climate, the Kids Are All Wrong," in which he referred to student activists as "a band of ignorant brats."[4])

In 2015, the Independent Petroleum Association of America (IPAA), a fossil fuel trade association, launched DivestmentFacts.com, which showcases some of these experts' arguments against divestment, as well as funds research. Its first report on "Fossil Fuel Divestment: A Costly and Ineffective Investment Strategy" was written by University of Chicago professor emeritus Daniel Fischel. The group also funded Professor Bradford Cornell (identified as a "visiting professor of financial economics at Caltech"[5]) to write a report showing divestment would lead to major financial losses.

Professors are not the only valuable pushback against student antics. A student at NYU's School of Professional and Continuing Studies, Eli Nachmany, wrote a letter in May 2015 to *The New York Times* arguing that divestment would mean significant losses to students, "including millions of dollars lost in available scholarships and grants."

How can university faculty and students be engaged or amplified to counteract threatening policies?

CHAPTER 3

RECRUITING UNIVERSITY EXPERTS

University scientists and scholars are needed to help set the record straight in scientific research, conferences, public communication, and courtroom testimonies. Alongside the military, scientists remain among the most trusted voices in society, with much greater trust than business executives, the national government, or the news media.[1] Their counterarguments are important proof that the science is not settled and that there is no scientific consensus and that therefore any regulation would be hasty. The following text can be copied or modified and used to gauge the interest and suitability of university scientists and scholars.

SO YOU WANT TO WORK FOR INDUSTRY?

You earned your PhD, then found a rare university faculty job, and maybe even succeeded in getting tenure and its associated job security. Along the way, you perhaps grew weary of the political views and media coverage of your more sanctimonious or famous colleagues. Maybe you see yourself as a "myth buster" who derives satisfaction in questioning scientific dogmas. Maybe you have become aware of the sprawling bureaucratic inefficiencies common in university life and have begun to sympathize with the private sector's predicament, which includes a refusal among people (especially your students) to accept personal responsibility for problems. Maybe you simply want to make some extra money.

As a university expert with such a disposition you may be uniquely suited for industry work. This is a guide to the benefits that you can expect, the characteristics industry is looking for, and the very minimal risks you face.

The benefits of being an industry expert are primarily monetary. Compensation for your efforts may come in the form of paid speaking engagements, contracts, consultancies, summer salary, grants, unrestricted gifts, research funds, travel funds, support for PhD students, ownership in company equity, and payment for expert testimony.[2] Some university experts have earned more from their courtroom testimonies on behalf of industry than from their university salaries.[3] It is not only the appeal of the amount of funding available, but also the relative ease of securing it. Many university professors find working with industry an easy way to grow their research profile with a steady stream of funding for graduate students and postdoctoral fellows.

These financial rewards will make you a more productive scholar and increase your status in your university, which champions researchers who bring in big money. Many universities have entire administrative units devoted to doting on deep-pocketed donors in the hopes of securing a large grant or gift agreement. They will be more than willing to help support your pursuit of industry funds.

While the university looks favorably upon these arrangements, your colleagues and mentors might be "incensed and disappointed" should you consult for industry.[4] In addition, a university expert is most valuable if the relationship to industry can be concealed or minimized so as to not compromise the patina of impartiality. For these reasons, you and/or the university may consider ways the industry can provide funds through alternative channels so as to avoid any red flags with university conflicts of interest offices. Or the industry may choose to bypass the university and instead pay you directly through consultancies, which most universities will not require you to disclose, at least not in terms of dollar amounts.[5] That way, the industry stays in the background and the university and its experts both continue to seem to be independent.

WHAT INDUSTRY LOOKS FOR IN AN EXPERT

In most cases, the Corporation, trade association, PR firm, or think tank will actively recruit university experts, although, occasionally, professors will approach industry. The economics faculty at George Mason University produced research consistent with the objectives of the Charles G. Koch Foundation (a charity spinoff of Koch Industries) and some have argued the professors there sought support from the Koch Foundation, rather than the other way around.[6]

Ideal experts have a "scientific pedigree, national standing, and propensity for public conflict."[7] An advanced degree from a respected institution is preferred, although it need not be in the area of expertise you may be asked to address. The ideal experts publish their research in high-profile scholarly journals.

If your views are consistent with the industry's—perhaps you share a preference for minimal state power (as did a number of prominent Cold War scientists who worked for industry)[8]—signal this shared sense of mission in an interview, an opinion piece, or on social media. This will help get the industry's attention.

If you love the product, show it. Cigarette manufacturers were keen to recruit experts who were also smokers.[9] Publicly defend the product in some way, perhaps by questioning the science that suggests the

Corporation's product is associated with a particular problem. If you want a contract from Facebook, it helps to be a Facebook user. If you are an animal scientist looking to work for the meat lobby, post photos of sizzling steaks on your social media account. The tobacco industry sought out scientists who questioned the link between cigarettes and cancer, or challenged the statistical methods, or presented competing theories on the causes of lung cancer. Not long after Dr. James Enstrom (who held a PhD in physics from Stanford but not a tenured university position) published work questioning whether low rates of cancer in Mormons were really related to them not using tobacco, the Tobacco Institute (a trade association) invited Dr. Enstrom to apply for a grant.[10]

The industry is looking for trusted voices, a profile that can change over time. The fossil fuel industry first recruited older, white males who often had PhDs in physics.[11] They fit the expert profile of that era, and were used across several sectors and issues, such as smoking and acid rain. But new faces are always needed. In the late 1990s, the American Petroleum Institute, a trade association for the energy sector, recognized they should recruit scientists "who do not have a long history of visibility and/or participation in the climate change debate," in part because some journalists were starting to connect existing contrarians to the energy industry.[12]

The profile of an "expert" changes as society changes. Professor Calestous Juma, who has helped Monsanto promote GMOs, is not only affiliated with the Harvard Kennedy School, but he is originally from Kenya and can offer a unique international perspective.[13] Dr. Sara Place, until recently the senior director of Sustainable Beef Production Research with the National Cattleman's Beef Association (now a consultant), is a rare woman expert among cattlemen.

Dr. Place is also skilled at using social media to defend the beef industry against climate-related accusations, as is her former mentor in the Department of Animal Science at the University of California, Davis, Dr. Frank Mitloehner. He addressed U.S. representative Alexandria Ocasio-Cortez directly on Twitter to make her aware that meat and milk represent a tiny fraction of U.S. emissions and argued the policy focus of the Green New Deal should be on transportation and energy production and use (not on cattle). In media coverage of the topic, Mitloehner gave Ocasio-Cortez "credit for reaching out" and himself credit for helping "set the record straight."[14]

There are also no cumbersome requirements to disclose funding in many forms of mass communication, which is where you will probably do the majority of your work for industry. Give your scientific opinion to reporters, on social media, and through Wikipedia. Given the right kind of expertise, you might be supported by Monsanto to post on social media, to create a website "testing popular claims against peer-reviewed science," or to write entries on WebMD about GMO labeling without ever mentioning your connection to the agrochemical industry.[15] Does $280,000 sound like decent compensation for a university professor to create a website "communicating the real story on fisheries sustainability" on behalf of the seafood industry?[16] You might even add to your personal website a page on "myths," where you dispute popular claims about overfishing and provide important "fact checks."[17]

DEGREES AND AFFILIATIONS

Mention that you have a PhD (on books, social media profiles, email signatures). Take any opportunity to bring up any accolades or scientific prizes. Always lead with your university title. Use it in research, talks, and public introductions. The more prestigious your institutional affiliation is, the better. For those without tenured positions at a university, any adjunct teaching can be used in your byline or in biographies. It is one thing for a company to challenge a study and call it "biased, flawed, and inaccurate" but it is much more powerful if someone perceived as an independent university expert makes that assessment. That is how Dr. Carl Seltzer, part-time staff for the Peabody Museum at Harvard University, described an early medical study linking smoking and heart disease. The tobacco industry recruited him to do research, and often referred to him as a "Harvard scientist."[18]

Whoever said that retirement is not the end of the road but the beginning of an open highway understood the value of a title that is retained after retirement. Emeritus professors are no longer subject to university oversight or social sanctions, yet still have access to the university's unique brand of authority. Put your emeritus title to work.

BE WILLING TO WORK OUTSIDE YOUR AREA OF EXPERTISE

Be prepared to engage with topics and methods that lie outside your formal training and to be presented as an expert in areas where you are not. "Harvard scientist" Dr. Seltzer had a PhD in physical anthropology but worked on smoking and heart disease. In a coauthored *Wall Street Journal* editorial in 2011 titled "The Myth of Killer Mercury," Dr. Willie Soon argued against a proposed policy that would require power plants to reduce mercury emissions. His byline described him as "a natural scientist at Harvard" and "an expert on mercury and public health issues." Dr. Soon is not faculty at Harvard University, but at the Harvard-Smithsonian Center for Astrophysics operated by Harvard College Observatory and Smithsonian Astrophysical Observatory, which is not a typical academic home for an expert on mercury and public health. Dr. Soon does not publish scientific research on mercury (his training was in astrophysics and engineering) but demonstrates admirable comfort with stretching his expertise. In a similar fashion, UC Davis's Dr. Mitloehner was not trained in climate science nor is he an "air quality expert" (which the dairy industry claims).[19] Dr. Mitloehner's PhD is in Animal Science from Texas Tech University, but that has not stopped him from effectively challenging scientific studies that implicate cows in climate change, nor from using the Twitter handle @GHGGuru.[20]

Nonexpert experts have an extra advantage when providing courtroom testimony because if they are "asked a potentially embarrassing question, this will be beyond their scope of competence" and they can refrain from comment. Many of the thirty-plus historians who have provided expert testimony had never published on tobacco and public health before they testified.[21] Chemical companies hired agronomist and father of the "Green Revolution" Dr. Norman Borlaug to advocate against regulating DDT, a pesticide that he did not study. He spoke at the 1972 EPA hearings about DDT, and Montrose Chemical Company, the largest U.S. manufacturer of DDT, arranged a successful press conference with him.[22]

BE WILLING TO COMPROMISE YOUR "ACADEMIC FREEDOM"

Do not cling to anachronistic notions of "academic freedom." In the 1930s, researchers who tested the effects of asbestos on animals commonly agreed that the experimental results were the property of the funders—in their case, the asbestos companies.[23] The ideal arrangement is one in which the industry owns the data you collect and can help frame the research results, although universities are making such an arrangement more difficult. Today, if you do wind up with data that make industry uncomfortable but are not exclusively within their control, be open to making a deal, including perhaps selling the data.

Consider signing a nonpublication agreement, nondisclosure agreement, or confidentiality clause that says you will not publish results, or that the funder has "the right to suppress the results."[24] Agree not to identify the company as a funder without the company's permission.[25] Sometimes the industry might stipulate the last 10 percent of funding contingent on agreeable results.[26] You might be able to publish, but required to consult with the funder about results or the publication date.[27] Some journals insist that investigators have the right to publish their results without control or consent of the funder,[28] but you can always find another journal with fewer restrictions. You might be asked "to return or destroy corporate documents" that you receive as part of a research project.[29] Allow the funder to see, comment on, and request changes to any manuscript before it is published. In the ideal scenario, you are a trustworthy ally and the industry does not have to make these requests explicit in contractual agreements.

DELIVERABLES

As an industry expert, your research is essential for defending against the risks posed by scientific knowledge. At the very least, frequently note that the problem is complex and more research is needed. Better still is if your research or opinion suggests that the problem is not a problem. The fishing industry might hire a university expert to testify before Congress about how the problem twenty years ago might have been overfishing, but today the problem is *under*fishing.[30]

Look for alternative causation. A soda company may hire experts to look for causes of obesity other than diet.[31] The tobacco industry might hire experts to look into the role of genes or stress (instead of smoking) as causes of cancer. A chemical manufacturer might be happy to fund a study that finds that pesticides are just one of "many factors that can contribute to failing honey bee colony health."[32] Unlike your colleagues, you prefer not to be hasty and believe in exploring every possible angle. The complexity of these issues means it will take a long time to get the facts straight, and by extension shows the public that they are unlikely to comprehend that complexity.

Research that shifts the spotlight of responsibility onto consumers, rather than producers, is desirable. You might propose research to study "addictive personalities" or other psychiatric features that predispose people to consume products. Perhaps you are willing to work with the alcohol, gambling, or tobacco industry to insist that the harms of addiction are limited to a small group of addicts.[33] Just do not propose to study any features of the actual products (such as marketing or affordability).[34]

If you find an issue with your research design, start over. If you realize you did not properly randomize your samples, rerun the experiment. If there is any doubt about any variable, rerun the experiment. Rerun them even if there is not. You can never be too sure. Meanwhile, the industry can assure regulatory agencies that someone is researching this question. Results just take time.

Be willing to point out "flaws" or "myths" in the existing scientific work or "debunk" it altogether. Show that those attacking the industry are "fraudulent" or represent "scientific misconduct" or promote "deliberate inaccuracies." Declare that research that points to the problem is too new to make any regulatory decisions, or, if the context demands it, claim those same findings are old news and have not necessitated regulation so far, and still should not.

Be comfortable with inconsistency and even hypocrisy. Promote your own findings, even if they have not been published in a peer-reviewed journal, but criticize scientists whose work threatens industry if they do the same thing. Claim that your model is the best all-purpose model, but, if necessary, point out that no model is perfect. Insist that animal studies cannot be used to make useful inferences about humans, unless the results support your position and in that case claim that animal studies provide the best available evidence (humans are animals, after all). If it serves your argument, claim the scientific standard for statistical significance (a p-value of less than 0.05) should be stricter. Or more lenient. Whichever is better.

CLAIM YOUR INDEPENDENCE

Your "independent views" are just that—beyond the scope of bias. Be unaware, skeptical, or dismissive of evidence that suggests financial interests consciously and unconsciously influence decisions at all points in the scientific process, including selecting a problem, formulating a hypothesis, designing an experiment, choosing your sample, collecting data, analyzing data, as well as interpreting and communicating results.[35]

Express complete confidence in your own incorruptibility. Dr. Richard Mattes, a full professor in nutrition science at Purdue University who received support from at least nine industry groups, including PepsiCo., responded to investigative reporting: "The science is what matters, not who funded the work. I have a very easy test for avoiding conflicts of interest. If I would hesitate to speak my mind about the outcome of honest science, I don't accept funding, engage in the work or publish the findings. Ethics starts with the individual, not the funding source."[36] Like the decision of whether to smoke, the question of ethics is a personal one.

The better a job you do on behalf of industry, the greater your future opportunities will be. That fact should not stop you from confidently declaring that you are "happy to be involved in the pursuit for truth."[37] Make statements about "the complete independence" made possible by industry funding.[38] You might declare "the funder had no role in study design, data collection and analysis, decision to publish, or preparation of the manuscript" in the acknowledgments section of the published study even if you did in fact have detailed email exchanges about the study's design with an industry representative.[39]

Editors, journalists, and your colleagues do not really know what a conflict of interest or bias is, either. Scientists are "gullible," and can be convinced that "any funding is a conflict of interest, even when the funder could not benefit in any way from the findings of the research."[40] This confusion will lead some to refer to "nonfinancial" or "emotional" or "ideological" conflicts of interest. "Conflation of 'conflicts of interest' with 'interests' in general serves to muddy the waters about how to manage conflicts of interest, generating confusion as to the nature and definition of the problem and doubt as to whether conflicts of interest can be addressed at all."[41]

You may even find opportunities to use the term "conflicts of interest" to your advantage. Friend of the beef industry Dr. Mitloehner tweeted on January 23, 2019: "This is Oxford's Dr Marco Springmann, the scientist behind much of the environmental portion of EAT Lancet [report on sustainable diets]. Some are concerned about biases and so am I. Why is an activist vegan not considered biased but a cattle rancher is?" He was not implying that Springmann financially benefits from veganism, but Dr. Mitloehner still was nonetheless able to use the occasion to create further confusion about different forms of bias.

Claim that receiving funds from industry is not any different from getting them from a foundation or the government. Most people will not understand that grants from industry are approved by the Corporation's legal departments to ensure they meet industry-friendly criteria.[42] You will benefit from your colleagues' confusion about whether giving time to an organization (such as serving on the board of a nonprofit organization) is equivalent to receiving money from industry. They will not typically understand the differences between being on the board of a charity (which usually comes with no pay) and the board of a company (which might include $50,000 or more a year).[43] Not only are the financial rewards different, so, too, are the missions of the institutions. Nevertheless, these assumptions about equivalency are to your advantage.

Speak about your industry funding in relative terms, which will help obscure the actual amounts of industry funding you have received. Hesitate to admit that the vast majority of your support comes directly from industry. Note instead something like that "the vast majority of [your] research funding comes from foundations"[44] (even if your foundation funding is not really the vast majority and those foundations might not have funded your research had they known about your ties to industry). When compared to your other sources of funding, direct industry funding represents "such a small percentage that it is inconsequential."[45] Prepare for one of your more loathsome colleagues to ask, if that is the case, why you would take it. Nobody will notice the paradox in justifying your industry funding on the grounds that industry should pay for research, while also saying that your industry funding is inconsequential and represents a very small percentage of your overall research funds.

HOW TO DISCUSS YOUR RELATIONSHIP TO INDUSTRY

For a long time, there was no reason to talk about your relationship with industry at all. Universities received industry funds quietly and without public notice. The extent of tobacco industry funding to universities is known not because experts disclosed their funding, but because of lawsuits and whistleblowers. Then some high-profile scandals in medicine in the 1980s initiated some action after fears that "professional judgment concerning a primary interest (such as a patient's welfare or the validity of research) [was] unduly influenced by a secondary interest (such as financial gain)"—including gifts to doctors and hospitals from pharmaceutical companies.[46]

In the name of protecting the "objectivity" of scientific knowledge, some universities, funding agencies, and scholarly journals now require conflicts of interest statements, which may include outside salary, speaker's fees, honoraria, and paid advisory positions.[47]

Some universities and government agencies began limiting conflicts of interest, including the time faculty could spend on or money faculty could receive from outside projects. In 2005, the International Agency for Research on Cancer announced that "scientists with real or apparent conflicts of interest" would no longer serve on the panels that produce important reports on the causes of cancer. A university-wide policy on Individual Financial Conflicts of Interest was implemented at Harvard University in 2010. More recently, the U.S. National Institutes of Health[48] and the U.K. National Health Service[49] put rules in place to address federally funded experts and their conflicts of interest.[50] Some universities began refusing money from tobacco and fossil fuel companies, likely in part to appease their more sanctimonious students.

But the rules around conflicts of interest and financial disclosure are hazy and inconsistent. At public U.S. universities, the only data reporters and activists can request access to are your formal grants and declared consultancies (but not contract amounts, fortunately) and often not even those details will be disclosed.[51] Dealings at private U.S. universities are inaccessible via public records requests, which means that researchers at private universities are even more attractive to industry. Some radical professors favor policies that would require full disclosure of funding for all faculty on university websites. Just what the university needs: another rule for paper pushers to implement.

Scientific journals also have varying policies. The competing interest policy at the scientific journal *Nature* recognizes that "it is difficult to specify a threshold at which a financial interest becomes significant, but note that many US universities require faculty members to disclose interests exceeding $10,000."[52] Some medical journals now refuse to publish tobacco-funded research,[53] but it is rare to see that prejudice leveraged against other industries. At least one high-profile journal[54] requests you declare "personal convictions (political, religious, ideological, or other) related to a paper's topic," an incoherent requirement that helps shift focus away from financial conflicts of interest. Many journals still have no competing interest policy at all, and no clear consequences for failure to comply with existing policies.

There are also no strict policies or norms for disclosing funding in press releases, newspaper opinion pieces, or on websites or social media. Even major newspapers such as *The New York Times* have at best a muddled conflict of interest policy.[55] Reporters regularly publish articles every year about how opinion authors fail to disclose their ties to companies. Nondisclosure is the norm. Take advantage of it.

IF INDUSTRY TIES COME TO LIGHT

You face a very minimal chance of anyone discovering your relationship to industry because you are likely to be the only person reflecting on your "competing interests."[56] On rare occasions, however, civil society groups, reporters, and colleagues may go snooping.

If you are called out for failing to declare your industry ties, do not panic. There are no formal punishments for scientific denial,[57] and no clear ramifications for nondisclosure. Your university, which has probably directly gained due to overhead from your industry funds, will almost certainly back you up. Consequences will be mild to nonexistent, especially when compared to those for other transgressions, like plagiarism, data fabrication, or sexual harassment, for which professors have lost their jobs. The Wikipedia entry for "scientific misconduct" does not even include the issue of disclosure. Undisclosed relationships have been discovered between university experts and all sorts of industries, including pharmaceutical companies, tobacco, fossil fuels, chemical manufacturers, food companies, the seafood industry, and on and on. Never once has it ended a career.

If you are called out for failing to disclose industry funding in a specific research article, you may be able to use the scholarly journal as a scapegoat. Emory University professor of psychiatry Dr. Charles Nemeroff was found to have earned hundreds of thousands of dollars giving talks about the antidepressant Paxil, sold by GlaxoSmithKline, while also managing a multimillion-dollar federal grant to do research on drugs made by the same company. "I have always been totally compliant, probably gone overboard, with disclosure," Dr. Nemeroff told a reporter who questioned his undisclosed financial ties. "If there is a fault here, it is with the journal's policy."[58] (Dr. Nemeroff left Emory after the scandal for a position at the University of Miami.) After Greenpeace showed that Dr. Willie Soon had not disclosed his funding from the energy utility Southern Company in a journal article that described dangerous climate change as a "flawed notion" that was "too speculative," Dr. Soon could hardly be blamed. The journal where the research was published, *Ecology Law Currents,* did not have a conflict of interest policy[59] and still does not. Many experts have "not felt obligated to disclose industry funding unless it was specifically for the research that is the focus of an academic journal article."[60]

When the failure to disclose industry ties is brought up for a specific research article, most journals wind up doing months of investigation and often simply stand by the nondisclosure. Occasionally, the journal may require an updated competing interests statement from the authors but, by that point, the study has already been covered by the media and influenced the public conversation. A study published in 2015 on hydraulic fracking "found no statistically significant relationship between dissolved methane concentrations in groundwater from domestic water wells and proximity to pre-existing oil or gas wells."[61] A reporter noted that several authors had not reported their ties to the Chesapeake Energy company.[62] Five weeks later, the journal published a correction in which the authors were keen to point out that "none of the authors have competing corporate financial interests exceeding guidelines presented by [the journal]."[63]

Regardless of how or when (if ever) your industry ties are brought to light, you can always downplay the influence industry funding has had on your research. Say the money was only for travel. Say it does not affect the results. After his funding from Monsanto was revealed, Dr. Kevin Folta of the University of Florida said, "It makes me really sad because I just want technology to help people. I don't even care about these companies. I want people to understand the science."[64] Declare that industry funding simply helps amplify the message, it does not influence the message itself. "What industry does is when they find people saying things they like, they make it possible for your voice to be heard in more places and more loudly," said University of Illinois professor emeritus Dr. Bruce Chassy, also with ties to Monsanto.[65]

Yes, you take money from industry but you would take money from anyone. Your expertise is in high demand by all sorts of different institutions. A reporter alleged that the restaurant industry paid an obesity expert, Dr. David Allison, then a professor at the University of Alabama, Birmingham, to write a brief for its legal case. The restaurant industry was challenging proposed regulations that would require chain restaurants in New York City to list the calorie content alongside menu items. Dr. Allison argued customers would indeed choose lower-calorie items, but they would go away hungry and overeat later to make up for it. He never disclosed to the reporter how much money he received from industry, but instead expressed gratitude to any funder who finds his opinion valuable. "Sometimes, when I'm involved in the pursuit for truth, I'm hired by the Federal Trade Commission. Sometimes I help them. Sometimes I help a group like the restaurant industry. I'm honored that people think my opinion is sufficiently valued and expert."[66]

If reporters or activists continue to attack you for your industry funding, say it is because they cannot attack the science. Note that investigations into your industry connections are "disturbing and problematic."[67] Just because someone does not like your results does not mean you are wrong. The science should "speak for itself."

Defend industry money by saying that of course industry should foot the bill for research.[68] Industry support is "a natural part of working on solutions."[69] It is not only appropriate but necessary to get to the bottom of things. If asked why, if industry funding is appropriate, you tried to hide it, say no one had previously asked.

You might occasionally be accused of being a puppet for the industry. During a television interview in 1970, Dr. Donald Spencer admitted he had not in fact written a leaflet that had been distributed by the National Agricultural Chemicals Association under his name, but that presumably a public relations firm had lifted the content "out of context" from talks he had given.[70] Try to keep your cool. When Dr. Folta was confronted with the fact that some pro-GMO text posted under his name online had been written by an employee at Monsanto, he simply promised he would write his own responses in the future.

You might wish to retaliate. After a *New York Times* reporter accused some professors, including Dr. Folta, of being in the pocket of the food industry, Dr. Folta filed a defamation suit against the newspaper (unfortunately, a federal judge in Florida threw the case out in 2019).[71] Dr. Folta remains gainfully employed at the University of Florida, where his teaching responsibilities include a course titled "Fruit for Fun and Profit."

University professors are almost never forced to resign over issues of nondisclosure. Even for those who hold research positions outside of universities, history shows there is another position waiting for you. In the fall of 2018, *The New York Times* reported that prominent breast cancer doctor at Memorial Sloan Kettering Cancer Center José Baselga had failed to disclose millions in industry funding from pharmaceutical and health care companies in dozens of his published scientific articles.[72] Soon after, Dr. Baselga stepped down from his position at Sloan Kettering and resigned from the board of the pharmaceutical company Bristol Myers Squibb. In less than four months, he was hired by AstraZeneca.

A WORD ON DEFECTION

Although it is very rare, some academic experts have expressed regret about their involvement with industry.[73] Do not cave to this feeling. Remember: "When it comes to regret, everyone's a winner!"[74]

If your conscience bothers you, do not try to assuage it by testifying against the industry. In 1975, a former medical officer for Johns Manville was going to provide testimony for an asbestos trial. At the last minute, the doctor withdrew because "he had received a telephone call 'from out West,' as he put it, warning him not to cross the Canadian border."[75]

Industry does not take kindly to turncoats. You may be followed, harassed, and disparaged. Consider what Syngenta did to University of California, Berkeley, professor of biology Tyrone Hayes.

In the 1990s, Hayes received funds from the consulting firm EcoRisk, funded in part by chemical manufacturer Syngenta and its predecessor, Novartis, to study the effects of the herbicide atrazine on frogs. Syngenta probably should have realized even at the time that Hayes was not sufficiently motivated by financial rewards, but he was just twenty-five years old, he had recently finished his PhD, his wife had gone back to school, and he had an infant son and no signs of family money. He seemed like someone who should have been motivated by the promise of reliable funding.

Hayes reported that in his experiment the atrazine-treated male frogs had smaller voice boxes, and that about one third of exposed males had malformations in their reproductive organs. These findings were bad for atrazine and for Syngenta, but Hayes had signed a confidentiality contract, which offered some protection. EcoRisk scientists suggested that Hayes run different statistical tests that would make the voice box effect go away. EcoRisk offered to buy his data from him. Hayes refused. Hayes wanted to present his initial results at a conference so he sent EcoRisk the abstract, and they delayed getting back to him so that he wound up missing the submission deadline. Hayes wanted to discuss his findings with an expert on the larynx and EcoRisk refused to agree to the discussion, which Hayes apparently found "personally offensive."[76]

Hayes disregarded industry advice and eventually turned against atrazine and Syngenta. He resigned from EcoRisk in 2000, and, in order not to legally defy his contractual obligations, he repeated his experiments and then published them.[77]

Because the usual course of incentives did not work to keep Hayes in check, an escalating set of disincentives was deployed. Syngenta hired a PR firm, which researched Hayes's personal life, wrote a psychological profile of him, and harassed him at scientific meetings. Industry representatives heckled him from the audience. Others made blatant threats. Yes, Syngenta's side proposed to "set a trap." True, they suggested investigating his wife.[78]

But, at every step of the way, Hayes refused to acquiesce. He used a study initially funded by industry against that same industry. He insisted on showing people photos of atrazine-treated frogs with no sperm in their testes despite the fact that statistics are the language of science, not photographs. He wrote to his students in 2002: "Science is a principle and a process of seeking truth. Truth cannot be purchased and, thus, truth cannot be altered by money."[79] He testified at government hearings and made all sorts of media appearances. He even started an anti-atrazine website called AtrazineLovers.com. Hayes sent Syngenta employees a flurry of taunting emails filled with "crude sexual innuendos and LL Cool J–inspired raps"[80] that Syngenta eventually posted online as proof that the professor was psychologically unhinged.[81]

The efforts to discredit Hayes initially seemed to work. His colleagues were patronizing. His former dean "told him that his communications have the potential to reflect poorly on him and diminish the impact of his research-based arguments." A reporter at *The New Yorker* even began looking into Hayes as someone who might be mentally unstable, yet capable of holding down a demanding faculty position. Hayes persisted.

Then someone went off script. Documents were leaked that showed that Syngenta had hired a PR firm to target Hayes in the various ways he had claimed. He was vindicated. After *The New Yorker* published the details in a 2014 profile, Hayes became a cause célèbre and an exemplary industry expert public relations nightmare. He is now so well-known and admired that he is effectively untouchable.

Few of you will be as incautious or defiant as Tyrone Hayes. You will run your experiments, rerun your experiments, build your models, ignore results that are unfavorable to industry, publish only those that are favorable, and disregard disclosure policies. You will publish your results. You will say there is no scientific consensus and present yourself as proof. You will provide helpful commentary on your social media accounts. You will cloak yourself in the superiority of the scientific method and its language. You will not make any rash judgments. You will claim to be apolitical. You will let industry stay in the shadows. And in return for your loyalty, your steadiness, and for staying on message, you will be rewarded.

CASE: HOW THE FOOD INDUSTRY COMMUNICATES ON OBESITY

The threat of intrusive food-related regulations at city, state, and federal levels has loomed large in the U.S. ever since the lead-up to the surgeon general's 2001 Call to Action to address obesity. Cities and states attempted to introduce a number of efforts to regulate foods in schools, nutrition labeling, and taxes on sugar-sweetened beverages. In 2010, the federal Affordable Care Act included a provision for menu labeling. These policy threats required restaurants and food companies to hire and create a vast network of defenders, including trade associations and the industry-backed Center for Consumer Freedom, whose mission is to push back against "a growing cabal of activists," which includes "self-anointed 'food police,' health campaigners, trial lawyers, personal-finance do-gooders, animal-rights misanthropes, and meddling bureaucrats" who all claim to know 'what's best for you.'"[1]

By outsourcing the pushback against policy, individual companies can claim they want to be part of the solution to public health issues. They can promote a new initiative or make declarations, like McDonald's did, that the company is "very responsive and responsible." Companies can stay in the background of policy debates, except to advocate for voluntary self-regulation.

In contrast to the individual companies, trade associations do the sensitive work of directly challenging the problem, the cause, the messengers, and the policies. In the years after the surgeon general's 2001 Call to Action on obesity, the Center for Consumer Freedom, often presented in media coverage as an independent organization, challenged the problem of obesity itself, and claimed that the problem was a "myth" or exaggerated. The National Soft Drink Association challenged the causal link between soda and obesity in children, with statements such as: "Childhood obesity is the result of many factors. Blaming it on a single factor, including soft drinks, is nutritional nonsense." The president of the National Restaurant Association argued that food establishments "should not be blamed for issues of personal responsibility and freedom of choice."[2] To a question about the role of restaurants in contributing to the obesity problem, he responded, "Just because we have electricity doesn't mean you have to electrocute yourself."[3]

How might these communications strategies be adapted to other issues?

CHAPTER 4

STRATEGIC COMMUNICATION

The risks of scientific knowledge are as much about the public's understanding of those risks as they are about the evidence for those risks. Therefore, the defense against scientific knowledge occurs on a battlefield of communications. This section outlines communications tips and tools.

There are few communication products that money cannot buy. Does the Corporation need a scientific journal sympathetic to its research? Fund an editor. Does the proper journal not exist? Create it. Establish a popular magazine. Build a website. Host a scientific gathering. Put together a group of people who appear to be grassroots activists. The digital media landscape in particular offers limitless options for shaping public perception.

Academic experts and consulting firms will communicate using scientific papers and the media. Think tanks in the Corporation's network will support experts to write books and help place highly visible editorials that summarize their findings. PR firms will assess how well experts are delivering their message and encourage them to craft a take-home message as part of publishing scholarly articles or speaking on television. They will also help shape how the Corporation's products are ranked in search engine results, including how to decouple search results so that queries about a problem do not return the Corporation and its products in the top results.[1] Polling agencies will conduct research to determine the target audiences for radio ads, television infomercials, mailers, and digital campaigns that challenge scientific knowledge or expertise.

STAY OUT FRONT

The number one rule of defensive strategic communication is to never be caught unaware. A team of consultants will monitor real-time press coverage for any scientific findings that threaten the Corporation. Make sure to "let no major unwarranted attack go unanswered" and to "make every effort to have an answer in the same day—not the next day or the next edition."[2] A single piece of research will rarely change the world, but the goal is to prevent individual studies from coalescing into a body of evidence. Challenge the research and its press coverage.

Insist that scientific work that threatens the Corporation is flawed. Say it on websites and social media platforms. Write to the journal to point out the study's weaknesses. Write a formal rebuttal. Whenever possible, fund a counterstudy.[3] Tell reporters they should not cover the work because the study has some issues and the journal is now investigating. By the time the conflict has been settled, the story will have gone cold.

Make scientists defend their results and their methods, and make journalists defend their sources. The day *The New York Times* published an article on atrazine,[4] Syngenta, the primary manufacturer of atrazine, planned to "go through the article line by line and find all 1) inaccuracies and 2) misrepresentations." Elizabeth Whelan, the president and founder of the American Council on

Science and Health (ACSH), who sought $100,000 from Syngenta that year in part for her attack on *The New York Times,*[5] gave an interview to MSNBC: "I'm a public-health professional," she said. "It really bothers me very much to see the *New York Times* front-page Sunday edition featuring an article about a bogus risk."[6] (Her performance was praised by Syngenta's vice president of corporate affairs.)[7]

Communicate with every generation. Help develop school curricula for all ages.[8] In 2014, Monsanto hired their first director of millennial engagement.[9] The U.S. beef industry has an "On the Farm" curriculum for schoolchildren of every age and, according to their website, more than fifty teachers used it in their classrooms in 2019.[10] In 2015, Dow Chemical pledged $1 million and founded the American Association of Chemistry Teachers.[11] The association's website describes the Science Coaches program, and highlights two middle school chemistry teachers who had students learn "about how and why GMOs are used."[12]

BUT DO NOT HESITATE TO IMITATE

The Corporation should also imitate and repurpose the tools of assault. Any characterization or tool that is at the disposal of activists, reporters, and even scientists is available to the Corporation and its arsenal, too. If scholars refer to the "echo chamber" of denial, begin to refer to the "echo chamber" of activism. If independent scientists use the term "competing interests" or "junk science" or "sound science," have industry scientists use those same terms.[13] If independent experts push back against industry-funded scientists, accuse them of "shooting the messenger."[14] Just as independent scientists and activists have called for industry-funded work to be retracted, call for the retraction of studies that are unfavorable. If reporters use government policies, such as the U.S. Freedom of Information Act, to obtain documents that show scientists have industry funding, use those same policies to harass university scientists who challenge industry.

The Corporation can modify any stamp of authority for its purposes as it challenges scientific knowledge. In response to the official IPCC—the Intergovernmental Panel on Climate Change—the Heartland Institute formed the NIPCC—the Nongovernmental International Panel on Climate Change. Industry-funded reports can be formatted to look like prestigious scientific journal articles. Any video can be made to look like a *60 Minutes* special. If the most popular climate activist is a teenager from Sweden, then hire, as the Heartland Institute did in 2020, another European teenager to be the "Anti-Greta" who pushes back against "climate alarmism."[15]

PRO-SCIENCE AND POLICY POSITIONING

Through it all, maintain that the Corporation is committed to being a force for social good. Express concern for public well-being and the health of the environment. The CEO of a deep-sea mining company can express earnestly that his efforts are "for the planet and the planet's children" and can hire a marketing firm to compare him to Elon Musk.[16] Express faith and optimism in human ingenuity and innovation. Emphasize the Corporation's importance to economic prosperity (while being careful nowadays not to brag about the Corporation's size).

No one has more reason to care about social issues than corporate executives. Remind the public that they are "parents, citizens, members of society" and this is a "perfectly honorable business" and that they "would not like to sell a product that was harmful."[17]

Support scientific conferences, create industry-sponsored scientific prizes, and support nonthreatening scientific research. The tobacco industry funded scientific research into basic health problems unrelated to tobacco use.[18] ExxonMobil gave money to top universities to support research on alternative energies.[19] In 2015, Monsanto announced a $4 million donation to monarch butterfly conservation efforts.[20] In very rare instances, express public support for a specific policy (especially if that policy is politically unrealistic). ExxonMobil has said it now supports a carbon tax. Facebook says it supports increased internet regulations.[21]

VICTIM POSITIONING

Avoid being portrayed as a villain. The Corporation is a hero, do-gooder, or, if necessary, the victim. If caught in a fraught situation, emphasize how unfair the treatment of the Corporation is and how dangerous this is to society. Remind everyone of the principle: innocent until proven guilty. Remind reporters that "it's easy to get stampeded, and the tobacco industry is being very much maligned. . . . Now the industry has been presented as a bunch of ogres trying to corrupt American youth," as did an executive from the advertising agency Benton & Bowles, at the time under contract of Philip Morris.[22] The tobacco manufacturer Brown & Williamson claimed cigarettes were being "brought to trial by lynch law."[23] "The demonization of carbon dioxide is just like the demonization of the poor Jews under Hitler," explained an industry-funded university expert.[24] After the Paris climate negotiations in 2015, the head coal lobbyist in Europe, who may have been confused about which side was victimized, said that the coal industry would be "hated and vilified in the same way that slave-traders were once hated and vilified."[25]

Refer to the Constitution if necessary. The Heartland Institute fought a subpoena that would have required the think tank to disclose any financial relationship with the chemical manufacturer Syngenta on the grounds it would constitute harassment and violate its First Amendment rights (the case was settled out of court in 2012).[26]

REINVENTION

Reinvention may be necessary. In 2000, negative press led British Petroleum to rebrand as Beyond Petroleum.[27] Trade associations can also change their names without changing their memberships or missions. CropLife America, which represents manufacturers of pesticides and other agricultural chemicals, was previously the American Crop Protection Association and, before that, the National Agricultural Chemicals Association.[28] After the Indoor Tanning Association was charged in 2010 with making a number of false claims[29] (including challenging the scientific evidence used in classifying UV radiation as a carcinogen),[30] tanning salons formed a new group in 2012 called the American Suntanning Association[31] ("a values-based organization dedicated to increasing public awareness about the facts associated with moderate UV exposure and spray-on tanning, correcting misinformation about sunlight and sunbeds").[32]

A trade association might need to be dissolved quickly in a crisis. In 2012, the U.S. Senate Finance Committee announced an investigation into manufacturers and advocates of prescription narcotics and that same day the American Pain Foundation disbanded, although they claimed the decision was made in advance of the investigation.[33] The U.S. Pain Foundation and the American Chronic Pain Association stepped into the void.

HAVE THIRD PARTIES TAKE THE DIFFICULT STANCES

In the past, corporate executives would respond firmly to allegations. In 1935, a reporter for *The Philadelphia Record* wrote a story about workers who had fallen ill at asbestos mills and accused the insurance industry of knowing about the hazards of asbestos since the late 1920s. The president of an asbestos company responded to the alleged cover-up in a letter to the editor and called it "unconfirmed dramatized tommyrot."[34] After an audience member at the 1996 shareholders' meeting of RJR Nabisco asked the company's chairman whether he would want people smoking around his children and grandchildren, he said that kids could leave the room if they were uncomfortable. When asked about infants, the chairman doubled down: "At some point they learn to crawl," a response that the record shows was met with laughter and applause.[35]

Such candidness is no longer advisable. CEOs should be sensitive. Instead, third parties, such as trade associations and consumer advocacy groups, can firmly defend the industry "without the risk of tarnishing the public image of individual companies."[36] Despite their partial funding by the food industry, trade associations and consumer advocacy groups funded by the Corporation are often described or alluded to in the news as independent organizations. This arrangement allows the Corporation and its CEO to stay in the background and assume the spotlight mainly to communicate positive, affirmative messages, while the trade associations do the dirty work.

COMMUNICATION PRODUCTS

With digital technologies, it has never been easier for the Corporation to pay to create, amplify, and control the scientific discussion. The digital media environment allows for websites, social media, and other forms of self-representation, which are cheaper and more ubiquitous. Gone are the days when the Tobacco Institute might mail out half a million copies of its newsletter *Tobacco and Health Report.*[37] The potential with digital media is many tens or hundreds of millions.

Sidestepping the hurdles of the peer review process associated with scientific advances has never been easier given digital tools. Press releases, advertising, and sponsored websites provide alternative platforms for challenging scientific knowledge. A study of more than 40,000 online articles shows that this tactic is effective: positions denying climate change were two times as prevalent on digital media as through conventional media, and more likely to feature climate change contrarians than scientists reflecting the consensus position.[38]

Websites

The Corporation can have websites created in an instant. The websites can be obviously related to industry, such as Vapor Voice ("a valuable resource for [the vaping] industry advances in regulation, legislation, product development, business profiles, scientific studies and other informative industry news")[39] or Beef. It's What's for Dinner ("discover the beef sustainability story from around the country"), or seemingly independent. Many domain services allow website owners to keep their personal information private, and there are no rules requiring websites to disclose their funding sources.

Industry websites can be used to comment on scientific work. The National Cattlemen's Beef Association can applaud a study on its website for confirming that the U.S. is the leader in sustainable beef and cattle production, but also note that it is "disappointed to see a statement in the abstract that advocated for a reduction in beef consumption" and call the statement "unfounded" and "inappropriate."[40] The Corporation can also pay experts to create websites or post content to third-party platforms that appears independent. In 2010, Monsanto sent an email to a University of Illinois professor with a plan for the professor to set up a website called Academics Review, which would publish expert responses to criticism of genetically modified crops and agrochemicals, such as glyphosate. Monsanto's chief of global scientific affairs said the problem could be "solved by paying experts to provide responses" but also that the "key will be keeping Monsanto in the background so as not to harm the credibility of the information." Monsanto has paid (directly or indirectly, through PR firms) bloggers and experts to write articles that support their chemicals and genetic modification on various websites, including WebMD.[41]

Search engine optimization

With the rise of the internet, a crucial component of communications has been search engine optimization. Purchase advertisement spaces for search terms of the names of individuals who have countermessages so that the Corporation's message can supplant the top results. For a time, an online search for Tyrone Hayes would generate a "Tyrone Hayes Not Credible" advertisement.[42] The month before reporter Carey Gillam's book about Monsanto was published, Monsanto's PR firm planned a step-by-step approach to ensure that online searches for her would return negative blogposts and websites. Previously, a Google search for "Exxon Secrets"—a Greenpeace campaign to expose ExxonMobil's climate change denial network—would return an advertisement for Exxon Energy Factor. A search for "meat tax" would return an advertisement that links to a National Cattlemen's Beef Association article on "Why Taxing Beef Isn't the Answer."[43]

Scientific journals

For the sake of scientific legitimacy, industry-funded research should continue to be published in scientific journals when possible. If there is not a sympathetic journal, create it. In the 1960s, the Tobacco Institute, a trade association, published the journal *Tobacco and Health Research*.[44] The Potato Association of American industry publishes the *American Journal of Potato Research* (where "100% of authors who answered a survey reported that they would definitely or probably publish in the journal again").[45]

There are also more subtle approaches, such as providing support to journal editors. The journals *Regulatory Toxicology and Pharmacology* and *Critical Reviews in Toxicology* have been supportive of the chemical manufacturers and their consulting firms.[46] When the team of sixteen scientists (ten of whom had consulted for Monsanto) wanted to challenge an international decision to list glyphosate as a carcinogen, they published their findings in *Critical Reviews in Toxicology*.[47]

Press releases

The importance of the press release cannot be overstated. Press releases can be used to promote scientific findings, respond to newsworthy events, and proactively shape the public conversation. Press releases for industry-funded research can emphasize different points than the original study and even take a stronger stance. A well-prepared PR firm has written standby press releases to defend the Corporation in almost any scenario, and has also prepared advice for executives about how to answer certain questions.[48]

Reporters are known to quote extensively or copy verbatim from press releases. A recent study of more than 1,700 press releases about climate change from various agencies over the last thirty years found that messages from business coalitions and trade associations that opposed climate policy were twice as likely to be covered as pro-climate policy positions from other types of organizations.[49] If the reason for the reprinting was merely reporter laziness, then *all* press releases would be reproduced by the media, but that was not the case. U.S. newspapers reprinted industry-backed messages opposing action on climate change around twice as often as messages from other sources that encourage climate action, showing not only the importance of press releases but also that the fossil fuel industry has a number of surreptitious allies in the media.

Op-eds

Op-eds are an opportunity to respond to real-time events or provide condensed versions of books. They capitalize on a current event or recent news. Syngenta, the main manufacturer of atrazine, hired the White House Writers Group (WHWG) to recruit scientists to "help them draft and place op-eds all over the farm areas of the target states" and brief national media including *The Wall Street Journal* and *Politico* on the atrazine issue.[50] An expert funded by the seafood industry wrote an opinion in *The New York Times* titled "Let Us Eat Fish" in which he claimed the scientific data on overfishing "were exaggerated." Op-eds can allow executives to speak to the public directly. In 2011, after environmental activists criticized destructive fishing practices associated with the tuna industry, the heads of three major tuna companies published an opinion in *The Wall Street Journal*. "Sustainable fisheries management is vital to our business, our employees, human nutrition and the planet's ecology. No one has more reason to keep tuna flourishing in the oceans than the people who depend on those tuna for their livelihoods," the executives wrote (or someone did on their behalf).[51]

Op-ads

A close imitation of the op-ed is an advertorial or an op-ad. This kind of advertising can promote not only the Corporation's product but also the Corporation's position for or against something (or someone) and can be published anywhere, including newspapers and scientific journals.[52] The advertorial "A Frank Statement to Cigarette Smokers" was published in 446 newspapers in 1954 and "may well be the most widely publicized—and expensive—single-page advertisement up to that point in human history."[53] It was a product of the PR firm Hill & Knowlton and underwritten explicitly by fourteen tobacco companies, and the text enumerated the ways in which "distinguished authorities" questioned the linkage between cigarette smoking and cancer.

Sometimes the line between advertorial and advertisement is thin. In 2008, Chevron, one of the largest fossil fuel producers, emphasized not only its product but its position on consumer responsibility with large posters featuring large-scale portraits with captions such as "I will unplug things more," "I will use less energy," and "I will take my golf clubs out of the trunk." Shell echoed this line of reasoning with a 2020 poll on its Twitter account asking, "What habits are individuals willing to give up to cut emissions?" (Significant blowback suggested that companies might want to avoid invoking individual responsibility on social media at this time.)[54]

Another, and less appreciated benefit of paid advertising is that the Corporation can threaten to withdraw advertising as leverage against the media. Several sponsors withdrew their sponsorship from the television network CBS after it aired a special segment on Rachel Carson.[55] Tobacco companies threatened to withhold advertising from magazines that gave too much coverage to tobacco and disease.[56]

Public letters, petitions, and pledges

Combat the illusion of "scientific consensus" by showing that large numbers of individuals disagree with mainstream science with letters to the editor, petitions, and pledges. The "Oregon Petition" was first circulated in 1998 and again in 2007 in response to the policy threat of the Kyoto Protocol. The petition argued there was "no convincing scientific evidence" of human-caused dangerous climate change and had more than 31,000 signatures, although more than 99 percent of signatories had no expertise in climate science (and some signatures, like Charles Darwin's, were forged). Nevertheless, the petition was an effective communications tool.[57]

Book reviews

Book reviews are a good way to mount a defense against a new book by a scientist or journalist. Both the Manufacturing Chemists' Association and National Agricultural Chemicals Association responded to Rachel Carson's *Silent Spring* with negative reviews.[58] Chemical corporations did the same to scientist Theo Colborn, who wrote about the effects of chemical exposure. In advance of the publication of Carey Gillam's 2017 book, *Whitewash: The Story of a Weed Killer, Cancer, and the Corruption of Science,* Monsanto's PR firm planned to "share links and information about how to post book reviews with industry & farmer customers" (perhaps the origin of a review of Gillam's book titled "Hogwash!").[59]

Speaking engagements

Scientists, journalists, and activists not only write about their work, they also speak about it, which provides opportunities to demonstrate there is scientific controversy. Syngenta and its PR firm Jayne Thompson & Associates tracked biologist Tyrone Hayes's speaking engagements and planned "systematic rebuttals of all TH [Tyrone Hayes] appearances."[60] A Syngenta representative was often there to "counter his outrageous accusations."[61]

Sometimes, however, it can be better not to engage and draw additional attention to anti-industry messages.[62] Efforts can be made to cancel public appearances and press conferences that pose a threat. After a medical researcher showed experimentally in the 1970s that inhalation of tobacco smoke led to tumors, the cigarette companies managed to cancel his press conference.[63] In 2016, a pro-Monsanto group effectively nixed an anti-GMO public lecture in Hawaii when they made more than 1,500 fake bookings for the talk.[64]

Museums

Brick-and-mortar opportunities to inform the public remain important, and museums are especially influential. The Corporation can sponsor museum content in independent institutions. In addition, the Corporation might consider educating the public directly through its own museum. The energy company Suncor in Alberta, Canada, underwrote the Oil Sands Discovery Centre located in Fort McMurray. The museum's website does not disclose its relationship to Suncor on its "About" page. Its mission is merely to "provide a unique and memorable experience that inspires a lifelong interest in science and technology." The focus at the museum is on technical details that emphasize the impressiveness of capital and engineering. The museum highlights "mighty machines" and the "recovery, extraction, upgrading, and reclamation" practices to shift attention away from competing perspectives, which have characterized the tar sands as "the largest and most destructive industrial project in human history."[65]

The part of the Oil Sands Discovery Centre dedicated to discussing the environment avoids the topics of water depletion and climate change, and focuses instead on beavers ("like beavers, humans can shape their environment to suit their needs, but they can also make environmental changes with far-reaching effects"). Suncor uses inclusive language to remind visitors that they, too, are part of the problem: "New research can help us avoid some problems and limit others, but we cannot use energy and have no impact at all. We share the benefits of this industry and all of us—oil sands companies, governments, and consumers—have a part to play in ensuring those impacts are sustainable." For their part, the visitors appeared satisfied with notes in the museum logbook that include "very informative" and "learned a lot!"

CASE: SOCIAL MEDIA "ADDICTION"?

A new technology emerges quite quickly and people begin to use it. Some people even use it a lot. Does that use constitute an addiction? After the invention of the bicycle, did individuals who rode their bike a lot get labeled as addicts? Were teenagers in the 1980s "addicted" to the telephone?

Excessive use of social media does not constitute "addiction," according to a number of experts. There is simply no solid evidence that people get addicted to social media. Preferable terms to describe the outcome are "binge watching" and "doomscrolling," which put the emphasis on the user doing the watching and the scrolling, not on how the platform promotes its content.

In any case, social media is just one of many things to be addicted to. If anything, the frequent use of social media is probably a symptom of an underlying problem, such as depression. More research is needed to say anything definitive. Just because depression and smartphone ownership are both on the rise does not mean the two are related. Correlation is not causation. In fact, social media use might actually be beneficial. Furthermore, individuals should have the right to regulate their own media intake.

What other ways might social media companies increase engagement, attribute responsibility for engagement to users, while at the same time fighting the stigma associated with encouraging addiction?

CHAPTER 5

CHALLENGE THE PROBLEM

The tools of science can be used to uncover human-caused or -exacerbated problems, such as a hole in the ozone layer, acid rain, or climate change. There are various ways to deny the problem. The options depend, at least to some extent, on features of the problem itself. Highly conspicuous health problems, such as obesity, are difficult to deny. More abstract problems, such as pesticide effects on wildlife or human hormones that are less visible, may be easier to deny. Climate change was a problem that fell somewhere in between. Fossil fuel companies adopted a position that climate change did not exist at all and, working together, held fast to the position, even after heat waves became more intense, droughts more common in certain regions, sea ice melted, and sea level rise accelerated, just as scientists predicted would happen.

HIDE OR DESTROY INTERNAL EVIDENCE OF THE PROBLEM

First and foremost, ensure that the Corporation's internal position regarding the existence of the problem is consistent with its outward-facing message. As early as the 1950s the fossil fuel industry was aware that their products were causing increases in atmospheric carbon dioxide concentrations. Some reporters and researchers now assert that ExxonMobil knew about climate change since at least 1977 (when it was just "Exxon") or even sooner.[1] Documents suggest an Exxon scientist spoke to Exxon's Management Committee in July 1977 and said "there is general scientific agreement that the most likely manner in which mankind is influencing the global climate is through carbon dioxide release from the burning of fossil fuels." In 1978, the same Exxon scientist said that there was "a time window of five to ten years before the need for hard decisions regarding changes in energy strategies might become critical."[2] These statements are now being presented as evidence against Exxon (and others) in climate-related litigation. Although ExxonMobil has responded that "ExxonKnew is a coordinated campaign perpetuated by activist groups with the aim of stigmatizing ExxonMobil" and has said the academic reports are "flawed" and have created a "false appearance that ExxonMobil has misrepresented its company research and investor disclosures on climate change to the public,"[3] it is obviously preferable not to leave any record of possible inconsistencies.

THERE IS NO PROBLEM

Deny the problem outright. The complete denial of a problem is a bold stance, but one that has been proven effective. The problem is a myth, a hoax, a fallacy, or "tommyrot."[4] Not only is there no problem, but attempts to convince the public there is a problem are propaganda. A coalition of fossil fuel companies formed a front group called the Information Council on the Environment (ICE) in the early 1990s to "reposition global warming as theory (not a fact)."[5]

Challenging the very existence of global warming as nonexistent stood in contrast to an earlier industry stance. Fossil fuel companies accepted that warming would indeed occur, but claimed humans would adapt (which rendered climate change real but a "nonproblem").[6] The approach adopted in the 1990s suggested there was no global warming *at all,* just natural variation. Exxon CEO Lee Raymond championed this position and claimed in the late 1990s that "in fact, the Earth is cooler today than it was twenty years ago."[7]

There is the opposite of a problem. Global warming is not a problem at all, but will mean better weather and better agricultural conditions. In 2003, *The New York Times* ran a story on fisheries bycatch—animals like fish, turtles, and sharks, and hundreds of thousands of dolphins and whales that are captured and thrown back into the sea dead each year. In the article, the National Fisheries Institute (the trade association for the American seafood industry) denied the problem of bycatch altogether and claimed that "some environmental groups are creating an aura of crisis, when in reality the opposite is true."[8]

Similarly, the NFL initially denied the problem of football-related concussions. When a team of medical researchers published the results in 2005 from an autopsy of a retired professional football player that showed brain damage from what was likely "repeated mild traumatic brain injury" from the sport,[9] the NFL hired medical experts who repeatedly called for the paper's retraction and pointed to "serious flaws" and "complete misunderstanding."[10]

IS THERE A PROBLEM? WE ARE LOOKING INTO IT

Did a scientific study suggest that there is a problem? Maybe there is, and maybe there isn't. Nobody knows more about the industry and cares more about worker safety or a sustainable, healthy environment than the Corporation. If there is a problem, it is very complex. There are a lot of unanswered questions. Industry-funded experts might ask, "But how do we determine whether all this new research is actually useful, unbiased and of high quality?"[11] The methodology is too new to be confident in the results. A meta-analysis may not even be "a scientifically valid method of analysis."[12] Why should models be trusted?[13] Exercise caution. Science takes time. That is why the Corporation is investing in research and committing to getting to the bottom of things (which delays regulation).

THERE IS A PROBLEM, BUT IT'S A SMALL PROBLEM

Minimize the problem by showing it is inconsequential or affects a very small number of people. Say that the problem has been exaggerated. Say the problem is small. Dr. Steve Koonin, a physicist, former chief scientist for BP, and currently director of the Center for Urban Science and Progress at New York University, published an op-ed in 2014 in *The Wall Street Journal* in which he noted that "human additions to carbon dioxide in the atmosphere by the middle of the 21st century are expected to directly shift the atmosphere's natural greenhouse effect by only 1% to 2%. Since the climate system is highly variable on its own, that smallness sets a very high bar for confidently projecting the consequences of human influences."[14] In addition to using minimizing language like "only 1% to 2%" and "smallness," Koonin focused on direct shifts and steered clear of the larger feedback effects from carbon dioxide leading to greater amounts of atmospheric water vapor—another potent greenhouse gas—and thus more warming (the earth system, however, cannot ignore an effect, whether it is direct or indirect).

The Corporation can point out that, more often than not, most people do not experience a problem. After factory workers showed signs of lead poisoning in the 1920s, the industry questioned why it was that other workers exposed to lead, such as chauffeurs, did not suffer.[15] As lead pollution grew, independent scientists estimated that lead levels were about 100 times higher due to industrial activities than they would have been naturally, and the lead industry pushed back that industrial lead pollution had only increased lead levels in the environment by a factor of two.[16]

THERE ARE BIGGER PROBLEMS

Call attention to the fact that there are bigger problems. DDT might kill birds, but malaria, which DDT helps prevent, kills people. The ACSH, largely funded by industry, published an article titled "Poverty, Not Climate Change, Remains World's Deadliest Problem."[17] A seafood-industry-funded scientist noted that even if there is overfishing, neither protected areas nor fisheries management will "shield marine biodiversity from the panoply of current threats: climate change and ocean acidification, land-based run-off, oil spills, plastics, ship traffic, tidal and wind farms, ocean mining and underwater communications cables."[18] Why focus on limiting fish catches with all these other problems in the oceans that need to be solved? Why address the livestock sector's contributions to climate change when transportation is a bigger contributor to emissions?[19] Deep-sea mining might harm the ocean environment, but other mining practices destroy rainforests and use child labor.[20]

THERE IS NO LONGER A PROBLEM

Perhaps there was once a problem, but now there is no longer a problem. Twenty years ago, overfishing might have been a problem but today "the US has largely 'solved' overfishing and has layered so much precaution into our system that the major threat to producing benefits is underutilization."[21] Concern about fishing in Australia is "outdated."[22] If there was a problem, it is now in the past.

CHANGE LANGUAGE TO ELIMINATE THE PROBLEM

Another way to make the problem go away is to change the language. The tobacco industry referred to "cancer" as "biological activity."[23] A consultant to the fossil fuel industry found that "climate change" sounded less frightening to a focus group than "global warming" and recommended the switch in 2002 (and the switch succeeded).[24] The chemical manufacturers insisted on the term "biosolids" instead of "toxic sludge."[25] The Ethyl Corporation convinced the Bureau of Mines to use the word "ethyl" in research reports instead of the term "tetraethyl lead."[26] Critics have said that the Marine Stewardship Council (MSC), a seafood eco-certification program, has certified overfished populations[27] but an

MSC employee told the journal *Nature,* "There are no overfished stocks carrying the MSC logo. They are all fished sustainably."[28] The MSC simply redefined the conventional scientific understanding of "overfished" as "depleted"[29]—one way to avoid granting their seal of approval to "overfished" fisheries.

CHANGE STATISTICS TO ELIMINATE THE PROBLEM

If it helps to eliminate the problem, use descriptive statistics instead of analytic statistics. Use a two-sided statistical test instead of a one-sided test. Argue that a different statistical test or statistical standard of significance should have been used. In 2015, U.S. government agencies announced that 2014 was the "warmest year ever." In a guest lecture at NYU that same year, Dr. Koonin (formerly of BP) criticized the U.S. government for being "dishonest" because the average temperature in 2014 was not statistically different from two previous years (even though 2014 was in fact, in absolute terms, the warmest year on record).[30]

CHANGE THE SCALE OF ANALYSIS TO MINIMIZE OR ELIMINATE THE PROBLEM

It is often possible to choose a particular scale of analysis—spatially, temporally, or in terms of sample size—to minimize or eliminate a problem. Ocean fish farmers minimize the problem of pollution (nitrogen and phosphorus from fish excretions) by narrowing their analysis to the area directly around the fish farm. Salmon farmers noted that "in deeper waters the effects of salmon farming are neutralized" and that "the pressure on the environment is kept under control when the current is moving all the activity around the farms."[31]

Insist that data used in scientific analyses meet certain criteria, which might mean the data come only from parts of the world that can afford to collect data to that standard. The seafood industry got its money's worth after an industry-funded expert teamed up with other scientists to respond to claims about global overfishing. The team looked at ten ecosystems and found that the average exploitation rate had declined in half of them. All ten of those ecosystems were chosen from the national waters of wealthy, developed countries that can afford detailed quantitative fisheries stock assessments and are best equipped to deal with management, but that did not stop the team from calling their study "global."[32] Researchers could similarly eliminate global poverty by restricting their analysis to Scandinavia.

Insisting on a certain standard of data can also prevent having to consider older evidence, which would reveal there was a problem. In that same study of "global" fisheries, scientists used assessments that started in the 1970s,[33] so they would not have to consider the full effects of industrial fishing, which began long before 1970. If the effects of fishing could be analyzed from 2015 onward, the problem of overfishing might be eliminated altogether.

Restrict sample size to what is best for business. At least one industry-funded study on the health of petroleum refinery workers did not categorize a contract worker (who cleaned reactor vessels) as a regular "exposed" employee, but did include a lawyer at the facility in the "exposed" sample.[34] The NFL hired a committee of physicians and researchers in 1994 who went on to publish thirteen peer-reviewed articles about how concussions did not cause long-term harm. A later investigation showed they came to these conclusions after eliminating more than 100 diagnosed concussions from their studies (by then, the research had helped buy more than a decade of policy delay).[35]

When advantageous, argue that subdividing a population is the *incorrect* scientific approach. In response to scientific claims that logging was causing declines in caribou populations, the Canadian forest industry insisted that caribou populations in Ontario were in fact "doing fine."[36] The Ontario Forest Industries Association noted in a report that "there are more caribou than deer, moose, and elk combined" and caribou are also "globally abundant." The problem was not a decline in caribou.[37] The problem was that the Canadian government chose to subdivide caribou populations into different "ecotypes"—caribou that live between the forest and the tundra and those that live in the woods. The forest industry pointed out, "Needless to say, if any animal population is subdivided enough times, the result is bound to be a very small population that can be considered at risk, threatened, or endangered."[38]

Other times it may be advantageous to argue that subdividing populations is the *correct* scientific approach. Bayer, a manufacturer of an insecticide that some scientific research had linked to honeybee population declines, used this tactic to challenge the problem of bee declines. A university expert with Bayer funding recommended that adding new beehives that have been subdivided from strong hives could be used when calculating hive loss rates, which would increase the denominator of total hives and lead to a reduction in the loss rate.[39]

The choice is obvious: choose the way to sort or express data that most benefits industry. Scientists can say that a wild fish population has "doubled"—even if that means an increase from merely 1 percent to 2 percent of their pre-fishing abundance.[40] Demanding greater precision from scientific analyses can also introduce uncertainty and buy time. There is always a genuine question about the best resolution for analyzing any activity. One group of scientists analyzed the presence or absence of fishing in 160,000 squares of oceans and found it was occurring in 55 percent of the oceans.[41] Another group of scientists, this one with seafood industry funding, challenged those results as "misleading."[42] They followed up with their own analysis, in which they divided the oceans into smaller squares (the size of a city block) and reported just 4 percent, not 55 percent, of the oceans were fished.[43] If the ocean as a whole was divided into brick-sized blocks, the percentage fished might decrease even further. Find the most favorable resolution or time frame for analysis.

THERE IS A PROBLEM, BUT PEOPLE ARE BETTER OFF NOT KNOWING ABOUT IT

In some cases, argue that people are often better off not knowing about the problem. A representative from a U.S. trade association for the insulation industry wrote a letter in the 1940s to a state official arguing against a proposed safety standard for products that contained asbestos, asserting that "this foreign disease . . . should be left in Europe" (by not speaking about it) and "not brought to our local communities and create hysteria and fear amongst the families of our contented workmen who are now enjoying good health and living to a ripe old age."[44] To prevent worker compensation claims, the asbestos manufacturer Johns Manville maintained a policy for decades of not telling employees if their physical examination showed signs of lung disease, and not sharing with sick employees the reason for their illness.[45] Contented workmen do not need to be made discontented. Eighty years later, a similar justification was provided for downplaying the coronavirus because it would only cause people and markets to panic.

THERE IS A PROBLEM, BUT IT'S NOT THE CORPORATION'S FAULT

If the Corporation must eventually acknowledge there is indeed a problem, such as global warming, cancer, or the collapse of honeybee populations, then there is often still a window of opportunity to deny that the Corporation has any role in causing that problem. Denial of causation is arguably the bread and butter of scientific denial and the subject of the next section.

CASE: QUESTIONING THE RELATIONSHIP BETWEEN VAPING AND COVID-19

In spring of 2020, a study showed that Chinese patients with Covid-19 were more than twice as likely to have severe infections if they smoked than if they did not.[1] As the virus spread, health officials urged people to quit smoking and also vaping because evidence suggested both smoking and e-cigarettes cause inflammation in the lungs, where Covid-19 often does the most harm.[2] Later that year, a team published a survey of 4,351 adolescents and young adults in the U.S. that indicated that smokers of both traditional and e-cigarettes were several times more likely to have received a Covid-19 diagnosis.[3] Highlighting that research alongside an individual case, *The New York Times* published an article, "Vaping Links to Covid Risk Are Becoming Clear."[4]

But that is just one side of the Covid and vaping story. In response to the research, seven experts sent a letter to the journal where the research was first published stating the "authors' conclusions are implausible based on the data presented in the paper" and "should not remain as is in the scientific literature and certainly not be featured prominently on your journal's website, as it is currently."[5] One of the seven signatories also coauthored an article, "COVID and Vaping: A Perfect Storm of Misleading Science and Media,"[6] in which she questioned "the relationship between COVID-19 and nicotine" and the "non-random sample" the study used, which was published at the online magazine *Filter* (there she also declared that she consulted for an e-cigarette company as well as a tobacco company).[7] Another article (written by someone with "unrestricted grants from tobacco manufacturers") on the *Tobacco Truth* blog also questioned the "minuscule case numbers" used to "make dramatic claims."[8]

Reflect on what public relations guru John W. Hill once said regarding tobacco use and lung cancer. "The public is in danger of being convinced that there is only one side. All the industry has a right to ask or expect now is for the public to understand that the case has not been proven and that there are 'two sides.'"[9] Discuss how this tactic can best be utilized to challenge causation.

CHAPTER 6

CHALLENGE CAUSATION

Should the Corporation publicly acknowledge that a problem exists, a popular and effective tactic is to deny scientific evidence that suggests a causal relationship between that problem and the Corporation's product or production methods. A study of the ten largest meat and dairy companies in the U.S. found evidence to suggest that each has worked to minimize the causal link between animal agriculture and climate change.[1] Again, challenging causation is the bread and, in that instance certainly, butter of buying time against costly regulations.

Scientific knowledge is at its most remarkable and perhaps most vulnerable when establishing a causal relationship. Scientific methods determined that giving baby aspirin to a child with chickenpox could increase their chances of Reye's syndrome—a dangerous swelling in the liver and brain.[2] Science is what led to the discovery that there are "short, critical times—like during fetal brain development—when chemicals can have disastrous impacts, even in very small concentrations."[3] The tools of science can help solve these mysteries, but when these discoveries come at the expense of industry, they must be challenged.

Scientists look for associations between variables and outcomes, but demonstrating causation can be difficult. The genuine difficulties and disagreements around establishing causation using the tools of science is a big part of the reason why denying the science of causation is possible and even, in some cases, easy. Many of the tactics used to deny causation will mirror those used to deny the problem itself, with some novelties. Sometimes the Corporation will want to challenge specific standards of evidence, and other times the Corporation will simply challenge causation in public communication by casting doubts on the findings using appeals to intuition and common sense.

If the challenge of causation is done correctly, the result is public perception that there is scientific controversy. The science is flawed or incomplete. There is science on both sides of the argument. Both sides. *Some* physicians question the cigarette-lung cancer link. *Some* scientists question climate change. *Some* experts question the contributions of cows to greenhouse gas emissions. The science is not settled when it comes to who is to blame. Funding counterstudies that question the causal relationship shows that the science is equivocal, which can delay policy.

HIDE OR DESTROY INTERNAL EVIDENCE OF CAUSATION

Again, ensure the Corporation's internal position and outward position are consistent. The suppression of internal research that shows any causal evidence is a good place to begin. If the Corporation has its own scientists who have found causal relationships between their products and harm, it will be necessary to suppress the evidence. Discourage employees from discussing these findings with outsiders. Label internal data as "not for public distribution," as General Motors labeled their own data from the early twentieth century about their cars and crash safety.[4] The chemical company DuPont suppressed experiments it had done on animals that showed a chemical called C8 used in Teflon coating was associated with health problems and even death.[5] (DuPont would have been wiser to destroy the studies altogether.)

THERE IS NO EVIDENCE OF CAUSATION

Remind the public that there is always a lot of uncertainty, certainly no consensus, and no conclusive scientific proof. When he was president of American Tobacco, Robert B. Walker said of cigarettes and lung cancer, "There isn't a mounting weight of evidence. There's a mounting wave of propaganda."[6]

Just because two things happen concurrently does not mean they are in fact related. Birdsong does not make spring come. Correlation is not causation. The "mere presence of a chemical in an ecosystem or ecological niche is by itself meaningless."[7]

The current science lacks rigor and clarity. "There is no rigorous (clinical trial) data on humans to show that red meat causes any kind of disease."[8] CropLife America, the trade association for manufacturers of pesticides and other agricultural chemicals, sponsored a television program that examined the decline of native bees and emphasized "experts aren't sure why" it was happening and "its causes are still unclear." From this uncertainty, the segment concluded rather certainly that "farmers, ranchers, [and] forest-land owners aren't the source of problems, they're the source of solutions"[9]—meaning farmers are not to blame for using pesticides.[10]

EVIDENCE OF CAUSATION IS WEAK, INSUFFICIENT, OR UNCERTAIN

If there is indeed some evidence of causation, remind the public that it is too early to say what the scientific findings represent. New research should be approached with caution. The evidence so far looks weak and insufficient. Scientific uncertainty can be emphasized to serve the Corporation's purpose.[11] "You should not act on hunches, suspicions, and stir-ups. This cancer business, now—nobody knows about it. I have to accept that there is some connection between smoking and health, but just what it is we don't know," said an executive from Benton & Bowles, an advertising agency for Philip Morris.[12] Some countries with high smoking rates have low lung cancer rates (the reasons for this paradox are not totally clear, and include aspects of smoking behavior, genetics, and diet).[13] Use these paradoxes to the Corporation's advantage to cast uncertainty over the pursuit of causation.

Fund a counterstudy to show that something, *anything,* about the original, implicating study is not true. When the U.N. Food and Agriculture Organization (FAO) published a report showing terrestrial animal agriculture emissions were equivalent to the transportation sector,[14] the Beef Checkoff program, an arm of U.S. beef producers, paid Dr. Mitloehner of the University of California, Davis, $26,000 to challenge the report[15] (although no funding was acknowledged in the published work).[16] Dr. Mitloehner did manage to find a problem, not with the animal emissions but with the transportation sector calculations (which had been used for comparison).[17] In the UC Davis press release, titled "Don't Blame Cows for Climate Change," Dr. Mitloehner called the FAO report a "lopsided analysis."[18] Rather than focusing on cows, he argued that the

FAO's "transportation analysis did not similarly add up emissions from well to wheel; instead, it considered only emissions from fossil fuels burned while driving." He did not challenge the absolute contribution of beef producers to climate change, but instead challenged the emissions beef produced relative to the transportation sector. The FAO admitted the transportation calculation was likely an underestimate, and Dr. Mitloehner's argument, which had not taken issue with the livestock calculation, nevertheless served to weaken the credibility of the FAO study.

APPEAL TO "NATURAL" PROPERTIES

The Corporation may appeal to the "natural" properties of the product itself and thereby appeal to common sense. How could this product possibly cause problems? How can carbon dioxide be bad? It is naturally present and makes plants grow. How can fossil fuels be bad? They are derived from once living organisms. The atmosphere is resilient against chlorine because it occurs naturally (via volcanoes, for one), so the stratosphere can certainly handle ozone-depleting chlorofluorocarbons (CFCs). Asbestos manufacturers believed that "asbestos was inherently useful, necessary, and therefore 'good.'"[19] How could eating meat be bad for anyone? Humans have canine teeth precisely because meat is a natural part of the human diet.

CHANGE STATISTICS TO ELIMINATE CAUSATION

Argue that the statistics used to demonstrate causation were incorrect. In earlier times, it was common to challenge the legitimacy of a statistical approach. "In all the uproar over whether there's any connection between smoking and lung cancer, one thing gets more and more clear—that you can prove anything you want with statistics."[20] Today, it is more common to say the statistical test is wrong or outdated than to challenge a statistical approach altogether. Challenge the specific statistical test or the appropriate p-value. Insist that the statistical significance is not significant enough. Whereas independent epidemiological experts have suggested that a 20 percent increase in risk of developing or dying of cancer is cause for concern,[21] experts working for tobacco argued that epidemiological studies related to secondhand smoke had to demonstrate at least 100 percent increase in risk to be credible.[22]

ANIMAL EXPERIMENTS AND CAUSATION

If experiments on animals serve the Corporation's purpose, then studies involving animals represent the highest standard of evidence. If they threaten the Corporation, challenge the relevance or validity of animal experiments. On the one hand, claim that "a two-year exhaustive examination of the effects of the plastic component BPA on rats might have finally put an end to the hysteria surrounding a chemical that has been used since the 1950s."[23] On the other hand, if activists start referring to scientific studies on rats showing that phthalates, used to make plastics soft, have caused cancer in lab tests, claim that humans are likely less sensitive to phthalates than rodents. "There is no reliable evidence that any phthalate, used as intended, has ever caused a health problem for a human."[24] Sometimes an experiment on rats can provide solid scientific evidence, other times rat studies are irrelevant.

CHANGE THE SCALE OF ANALYSIS TO MINIMIZE OR ELIMINATE CAUSATION

Change the scale of analysis to eliminate or minimize causation. When an international report attributed 18 percent of global greenhouse gas emissions to livestock, a counterstudy examined only the state of California and "arrived at much different GHG [greenhouse gas] estimates associated with direct livestock emissions (enteric fermentation and manure), totaling at less than 3% of total anthropogenic GHG and much smaller indirect emissions compared to the global assessment."[25] Narrow down the analysis by, for example, focusing on a single greenhouse gas. The animal agriculture industry and its allies will admit that animals do produce greenhouse gases, and then strategically focus on how they will address carbon dioxide, rather than methane, which accounts for a much larger share of emissions.[26]

SOMETHING ELSE CAUSES THE PROBLEM

Proposing a "credible alternative hypothesis" is an important technique. Changing the scale of analysis can create an alternative explanation for the cause of the problem. When a scientist wrote to Syngenta saying that his initial research was showing it "unlikely" that the varroa mite was responsible for the honeybee declines, an executive suggested various ways to tweak the study, including only analyzing changes in Europe. Voila: varroa mites, not pesticides, became a credible alternative cause for the decline in bees.[27] Lung cancer is the result of an individual's genetics or stress levels, not of the amount they smoke. Acid rain is not caused by sulfur dioxide pollution from humans, but from volcanoes like Mount St. Helens.[28] The sugar industry sees credible alternative hypotheses in "not only saturated fat, cholesterol, and salt, but also portion sizes, processed food, sleeping habits, lack of exercise, environmental toxins, viruses, prescription drugs, and even alterations to our microbiome."[29] To whatever extent global warming occurs, it is the result of natural variation. Fish populations are not declining because of overfishing, but due to warmer waters. If there are already scientists who have looked for alternative causation independently, amplify their research. If there are no scientific studies, fund them into existence.

THERE ARE BIGGER CAUSES OF THE PROBLEM

Point to bigger causes of the problem. The cause of lead poisoning in children is lead paint, not leaded gasoline.[30] The cause of global warming is not U.S. emissions, but the current and future emissions from China, India, and Brazil. Herbicides may play a role in the decline of the monarch butterflies, but it is minor compared to urban development, plant disease, and climate change. When it comes to climate change, the beef industry can shift the focus to sport utility vehicles. When it comes to human health, meat companies can remind the public of sugar and processed foods. Fish farmers say that household pets, not farmed fish, are responsible for the demand for fishmeal (which leads to the overexploitation of wild fish).

CONSUMERS ARE RESPONSIBLE FOR THE PROBLEM

If the Corporation must accept that its product causes a problem, then there is still the question of responsibility for its use. Remind people that without demand there would be no supply (ignore the aphorism that if Henry Ford had asked consumers what they wanted, they would have asked for a faster horse). The Corporation's product may cause a problem, but the consumer chose to use it.

The tobacco industry shifted responsibility away from the companies and onto customers.[31] Executives made statements like: "I don't think any industry should be persecuted for the immoderation of its users, provided the industry hasn't promoted immoderation—and certainly the tobacco industry hasn't."[32] BP helped popularize the concept of a "carbon footprint" and even introduced a "carbon footprint calculator" in 2004 that consumers could use to tally their emissions.[33] In a seminar on legal responsibility for climate change, a lawyer for ExxonMobil asked: "What about all of you in this? All of us? I drive a car. Why am I not named a defendant?"

Individual responsibility may not have to do with actual consumption, but with consumer behavior. Car manufacturers claimed injuries in car accidents were the result of bad drivers, not bad car design.[34] Food and beverage companies argue teenagers are not overweight because of junk food or soda, but because of their own lack of activity.

WORKERS ARE RESPONSIBLE FOR THE PROBLEM

Implicate workers in the cause of the problem. In the early twentieth century, the U.S. Radium Corporation said that dial painters (for watches) were not suited to more difficult labor and it was their inherent weakness, not their use of radioactive radium, that led to their poor health.[35] Lead manufacturers shifted the blame to workers, and said they had sloppy habits and were careless. The asbestos manufacturers could not help it if workers did not wear masks to protect themselves from asbestos hazards (never mind the fact that workers did not know about the hazards). The dangling ponytail of a trainer killed by a captive orca whale that defied protocols was to blame for her death, not SeaWorld.[36] When a pork processing plant in South Dakota became a coronavirus hotspot, Smithfield Foods blamed employee "living circumstances."[37]

THE GOVERNMENT IS RESPONSIBLE FOR FAILING TO PREVENT THE PROBLEM

When all else fails, blame the regulators for failing to regulate. As one former executive noted, "Manville acknowledged that [asbestos] was potentially harmful but insisted that employees, unions, customers, regulators, scientists, and insurance companies all knew of the dangers. . . . During World War II, for example, the U.S. government controlled the use and applications of asbestos as a strategic and critical war material. Surely the government should bear some responsibility for the ensuing problems."[38]

OUR PRODUCT CAUSES A PROBLEM, BUT IT IS SMALL AND CONTAINED

If the Corporation must acknowledge scientific evidence of causation, note that the problem is small and contained. Purdue regularly claimed that the risk of addiction from its painkiller OxyContin was "less than one percent."[39] Cigarette manufacturers often pointed out that many smokers did not get cancer. The NFL has reminded us that "there are thousands of retired [football] players who do not have memory problems."[40] The problems of alcohol abuse and gambling addiction are limited to a small number of individuals with addictive personalities. There is no need to regulate all of society.

WE CAUSE THE PROBLEM, BUT WE ARE GETTING BETTER

If the Corporation acknowledges some responsibility for causation, focus on how practices have improved, which implies no regulation is necessary. The beef industry has argued there have been "improvements in efficiency" and the carbon footprint of beef in 2007 was 16.3 percent lower than it was thirty years prior.[41] The U.S. dairy industry "has made remarkable productivity gains and environmental progress over time."[42]

THE BENEFITS OUTWEIGH THE RISKS

Concede that there may be some risks associated with the product, but they are outweighed by the benefits. Even if there are associated health effects, cigarettes give so much pleasure to so many people. In the 1960s, a representative from American Cyanamid Company, a leading manufacturer of agricultural chemicals, argued that birds were actually doing better than ever, even with pesticides and DDT, and that one could "make out a case for pesticides as being a major cause of bird survival through the reduction of insect-vectored avian disease."[43] On balance, pesticides are good for birds. Carbon dioxide has caused global warming, but it has also helped make plants grow, and those benefits "will easily outweigh any negative effects."

THERE ARE RISKS, AS WITH EVERYTHING

Make the point that life inevitably involves risks. One tobacco executive noted that "everything around us has an element of danger. Your swimming pool can kill you. Cars can kill you. Coffee can kill you."[44] "Safety is relative, not absolute."[45] "Our life in advanced societies has become so safe that people are outraged at any hint of a theoretical risk."[46] Risk, of this product or otherwise, is just a fact of life.

CASE: BIAS IN THE EAT-*LANCET* COMMISSION

The EAT-*Lancet* Commission led by researchers from the Harvard School of Public Health and Stockholm University "brought together 19 Commissioners and 18 co-authors from 16 countries in various fields including human health, agriculture, political science and environmental sustainability" to examine how to feed 10 billion people a healthy diet within the ecological boundaries of the planet.[1] Their report was published in January 2019 and highlighted a need to significantly reduce animal agriculture for reasons of both human health and the environment. According to an analysis of 8.5 million tweets, a large countermovement (identified by the hashtag #yes2meat) organized a week before the report's launch date and continued after the report was published.

Of the top twenty most shared online sources on Twitter, nine were neutral or positive about the report while eleven were critical of the study. All but one of the eleven most popular critical articles were published on alternative media platforms.[2] Two of the eleven critical articles were by Nina Teicholz—one on her own blog and the other published on the website of the Nutrition Coalition (where she is executive director). On her blog, Teicholz noted that "an examination of the EAT-*Lancet* authors reveals that more than 80% of them (31 out of 37) espoused vegetarian views *before* joining the EAT-*Lancet* project." She referred to the commission as "highly biased" and wrote they could not "be expected to produce a balanced outcome" with "like-minded people talking to themselves" and "inbred conversations."[3] Teicholz linked to that blogpost in her article for the Nutrition Coalition, where she called the report "one-sided" and suggested the commission was hypocritical. "One could ask, further, about the

GHG emitted by the whole EAT-*Lancet* project . . . ," she wrote. "How much GHG was required to enable all this travel?"[4] Another blogpost challenged the commission's legitimacy: "The idea that a group of privileged, wealthy, white, thin people can come up with one global dietary solution is insane."[5]

Which weaknesses and biases can be exploited in the scientists, reporters, activists, or lawyers whose efforts pose a risk to business operations?

CHAPTER 7

CHALLENGE THE MESSENGER

In defending the Corporation against an assault by the scientific establishment, it may be advisable to challenge some of the members of that establishment, including institutions, and also individual scientists, scholars, physicians, reporters, and activists. Before he died, molecular biologist Andrés Carrasco worked at the University of Buenos Aires studying the effects of the pesticide glyphosate on frog and chicken embryo development.[1] He described the "malformations" he observed and suggested there could be similar effects in humans. He became one of Monsanto's "most difficult public relations problems."[2] This section suggests how to deal with the Carrascos of the world.

Challenging individuals is a sensitive subject. When eight tobacco executives met at the Plaza Hotel in 1953, they agreed: "It was the unanimous feeling of the committee that under no circumstances [underlining in original] should the industry attack the integrity of the scientific researchers who have been leading the attack against cigarette smoking."[3] The president of R.J. Reynolds, Edward Darr, may have silently disapproved of this approach because he had already attacked the integrity of physicians when, a year earlier, he had accused them of compromising the truth in their pursuit of the limelight: "One of the best ways of getting publicity is for a doctor to make some startling claim relative to people's health regardless of whether such statement is based on fact or theory."[4]

Whatever opposition there was to attacking the integrity of experts would soon dissolve, possibly at the behest of PR professionals. In July 1954, a Hill & Knowlton staff economist laid out the options for handling antismoking experts to the president of Brown & Williamson Tobacco Corporation:

> (a) Smearing and belittling them;
> (b) Trying to overwhelm them with mass publication of the opposed viewpoints of other specialists;
> (c) Debating them in the public arena; or
> (d) We can determine to raise the issue far above them, so that they are hardly even mentioned; and then we can make our real case.[5]

Hill & Knowlton, trailblazers in so many ways, would use a combination of these options against the cadre of independent scientific experts that implicated cigarette manufacturers in public health issues. Today, attacking the integrity of the scientific researchers is a tried and trusted tactic.

Most of the targets of these attacks will have no experience in the trenches: no training, no familiarity with these rebuffs or how to handle them, and no legal counsel. They will be unprepared for the systematic rebuttals of both the work and those responsible for it. The predictable pushback in print, on social media, and at speaking engagements will discourage others from these lines of inquiry, referred to as "the chilling effect."

In some circumstances, the Corporation may be required to denigrate the notion of scientific institutions, scientific authority, and the credibility of the media. On behalf of the tobacco industry, Hill & Knowlton challenged public health organizations, including the American Heart Association. The fossil fuel industry, the agrochemical manufacturers, and others have emphasized the ineptitude of the EPA. The industry-funded ACSH regularly challenges the competence of the mainstream media. "If you want to reinforce your anti-biotechnology, pro-alternative medicine beliefs, read the *New York Times*. If you want serious science journalism, read something else. Nearly anything else. Preferably us, though."[6]

The Corporation may take the position as a defender of democracy (there is, after all, a fine line between government regulation and authoritarianism) and scientific integrity. In response to a U.S. surgeon general's report about secondhand smoke, released in 1986, the Tobacco Institute (the trade association and brainchild of the PR firm Hill & Knowlton) published its own report on secondhand smoke, with the subtitle "Scientific Integrity at the Crossroads," and publicly accused the surgeon general of "censorship and abuse of science."[7]

Finally, another result of the paid attacks on scientists, journalists, and activists has been the normalization of such attacks, which has energized citizens to participate in the same kinds of arguments, especially in comment sections of newspapers and on social media. This unpaid militia of defenders helps to spread the Corporation's message and take on a lot of the legal responsibility. They, too, have inevitably helped to dissuade scientists from pursuing controversial research.

DISCOURAGE INTERNAL DISSENT

It is important to make examples of company dissidents or "whistleblowers." At a company meeting in 2020, a technical expert for Exxon said he brought up the company's failures to address climate change and asked if this was "a problem of behaviors and leadership—not science and technology." He alleges that a few months later he was told he could either participate in a three-month remedial program that would require regular meetings with his supervisor and, if his performance was unsatisfactory, he could be fired, or he could resign.[8] In the 1990s, an employee for Koch Industries, Sally Barnes-Soliz, reported that the company had underreported its benzene emissions to Texas state regulators (they had produced fifteen times the legal limit). Koch Industries reportedly put Barnes-Soliz in an empty office with no responsibilities and no email access. She eventually quit (and sued Koch Industries for harassment; the case was settled in 1999).[9] Universities likewise do not respect "whistleblowers" who jeopardize large industry grants and they, too, have relegated rogue researchers to lonely offices with no responsibilities.[10]

Funding for any third-party experts who go off script should be terminated. In 1963, a scientist who had been receiving funding from the trade association API for nine years (totaling about $200,000) published a paper about how lead in the environment was mainly the result of the oil industry.[11] After that, the API stopped funding him.[12] In the 1970s, after a medical researcher funded by the Council for Tobacco Research (formerly the Tobacco Industry Research Committee) reproduced results that inhalation of tobacco smoke led to tumors, the council did not renew his contract. More recently, a scientist who publicly criticized deep seabed mining was allegedly warned he might lose his industry funding if he continued to speak out.[13]

OFFER AN INDUCEMENT

One way to dissuade a scientist or a reporter from continuing their line of research is to offer them an inducement. Geochemist and Caltech professor Clair Patterson, who studied lead pollution, said that the petroleum industry offered him funding to study something else.[14] Among the list of options chemical manufacturer Syngenta considered using against biologist Tyrone Hayes was offering "to cut him in on unlimited research funds."[15] While it did not work in these instances, there are an untold number of scientists and reporters for whom such an incentive can successfully refocus their work.

INVESTIGATE INDIVIDUALS

If positive incentives cannot achieve dissuasion, intimidation by investigation might. Private investigators and some public relations firms will proactively background-check scientists and reporters. Hiring a private investigator is also an option. Syngenta hired detectives to investigate scientist advisers to the government on the question of whether to regulate atrazine.[16] According to scientist Andrés Carrasco, the Chamber of Agricultural Health and Fertilizers (which represents Monsanto and the other agrochemical industry interests in Argentina) sent representatives to visit his laboratory to look for relevant research documents.[17] After reporter Paul Brodeur's series of articles about the asbestos industry was published in *The New Yorker* in the 1980s, Brodeur alleged he was followed.[18] After the publication of her book *Dark Money* and being followed by a private investigator, reporter Jane Mayer remarked, "I've been a reporter for a long time, covering wars, the CIA, presidencies and a lot of very powerful organizations. But the Kochs [David and Charles] are the only people I've ever covered who have hired a private investigator to try to dig up dirt and plant untrue stories about me in order to hurt my reputation." Like the Koch brothers, refuse to comment on such matters.[19]

LEGAL INTIMIDATION

Legal intimidation is also common, and was being used as early as the 1940s. The Lead Industries Association threatened to sue Randolph Byers, a pediatric neurologist and Harvard Medical School faculty member, after he coauthored a study suggesting that lead poisoning in children could lead to poor performance in school.[20] The tobacco industry has sued University of California professor Stanton Glantz, who conducts research on cigarettes and vaping, more than once.[21]

Subpoenas and depositions can also be used to intimidate scientists, reporters, and activists. R.J. Reynolds subpoenaed historian Robert Proctor's book manuscript on the tobacco industry before it was published (although judges intervened).[22] Monsanto subpoenaed emails from scientists who study glyphosate.[23] During the Deepwater Horizon oil spill, BP subpoenaed more than 3,000 emails from researchers at the Woods Hole Oceanographic Institution under the guise of verifying the accuracy of their research.[24] After he agreed to testify on behalf of the plaintiffs in the 2000s, the tobacco industry deposed historian Louis Kyriakoudes so many times that it got in the way of his research. He "cut back to one or two trials a year," noting that "harassment is effective"[25] (which explains why it is included here).

OTHER FORMS OF INTIMIDATION

While background investigations are intended to be covert, there are also overt and menacing forms of intimidation. After a 1993 study[26] showed that households with guns were in fact at higher rather than lower risk for homicide, the lead author reported he received notice from the National Rifle Association's Institute for Legislative Action that he now had his "own, named file at the NRA headquarters."[27]

In plenty of places, physical intimidation is routine. Sofia Gatica, an activist in Argentina working against agrochemicals, has received repeated death threats, including once at gunpoint when she was told to "stop messing with the soy."[28] At least 212 environmental activists were killed in 2019, and the deadliest issue, with a reported fifty deaths, was mining, followed by agribusiness.[29]

CLAIMS OF SCIENTIFIC MISCONDUCT

Formally accuse scientists, particularly those funded by federal grants, of scientific misconduct. Write to their universities, deans, or department heads.[30] The Ethyl Corporation pressed charges of misconduct in the early 1990s against Herbert Needleman, a medical researcher who had received funding from the National Institutes of Health for his scientific studies on lead and its effects on children. His data and analyses were then investigated by the University of Pittsburgh, where he worked. Needleman described the experience. "Horrible. It was absolutely horrible. I was so angry, and it's not good to be that angry and worried; it's bad for your health."[31] Bad for Needleman's health, perhaps, but good for discouraging others from pursuing similar research.

A scientist for the U.S. Department of Agriculture, which has strong ties to agrochemical manufacturers, told reporters that he was reprimanded at work after he did research that examined the risks of agrochemicals to pollinators. He claims he was told not to talk to the press "without prior approval." He was also accused of defying the paperwork protocols for travel and of office misconduct for various infractions, one of which was "dancing around the office and pretending to hump a chair." Three other scientists who work on pollinator research have also reportedly experienced retaliation.[32]

CLAIMS OF BIAS

The suggestion of "bias" can diminish credibility. Claim scientists and journalists are, in fact, activists. Insist that they lack the objectivity and neutrality that their professions demand. When *The Guardian* approached ExxonMobil for an interview in 2015, the company responded: "ExxonMobil will not respond to Guardian inquiries because of its lack of objectivity on climate change reporting demonstrated by its campaign against companies that provide energy necessary for modern life, including newspapers."[33] After a group of journalism postgraduate fellows at Columbia University published two articles about ExxonMobil's climate change research in the *Los Angeles Times* in 2015, ExxonMobil sent a letter to the university's president and board of trustees saying the reports were "deliberately misleading."[34] The agrochemical industry insisted that an international panel's decision that the herbicide glyphosate is a carcinogen should be dismissed because a scientist who participated (although he did not cast a vote on the decision) also did part-time consulting for an environmental organization.[35]

Scientists criticized for industry funding will often counter that scientists "on the other side" are funded by mission-driven nonprofits or philanthropists. A scientist funded by the beef industry has raised concerns about bias in a vegan scientist who coauthored a report about sustainable diets. Vegans are biased (meat eaters are just normal). For scientists who are not compensated for their work, accuse them of even greater bias because a willingness to work for free shows the individual must be personally invested in the research outcomes.

CLAIMS OF ALARMISM

Suggest the messenger is an alarmist, apocalyptic, a crusader, a cult member, a doom-and-gloomer, a fanatic, emotional, a fearmonger, hysterical, a killjoy, and, especially damning in the U.S:, a pessimist. Elizabeth Whelan (who wrote *Panic in the Pantry* in 1992 and founded the American Council on Science and Health) is credited with coining the term "chemophobia."[36] Monsanto-funded experts have referred to their opponents as "anti-GM chemophobes." Scientists funded by the seafood industry claim that consumers have been "misled by prophets of doom and gloom"[37] and that the media favor "apocalyptic rhetoric that obscures the true issues that fisheries face."[38]

CLAIMS OF BEING BORING

Maintain that the messengers are just saying the same old thing. This again? Emphasize how tired everyone is of blaming industry. This latest controversy is just one in a long line of controversies. The tobacco industry funded a report in the late 1960s titled "Centuries Old Smoking/Health Controversy Continues" that argued societies had been fighting over cigarette use since the seventeenth century.[39] The ACSH has paradoxically called topics alarmist and boring in the same breath. One article begins, "Bisphenol A—a long-used component of polycarbonate plastics, is one of the most studied chemicals in the world. Even the ultra-cautious FDA has declared it safe for people as used. But some scientists have built a career by screaming about how dangerous it is, so we have another paper. Enough already."[40] The ACSH also sometimes reminds readers just how boring the news they repackage is: "More of the same: Another slanted anti-pharma op-ed. Yawn."[41]

CLAIMS OF LYING

Accuse reporters and scientists of deception. The ACSH uses this tactic frequently in the web headlines against reporters who cover glyphosate, which Monsanto, an ACSH funder, makes, including "Glyphosate: NYT's Danny Hakim Is Lying to You" and "On Glyphosate, Carey Gillam Keeps Lying Early and Often." After Bayer and Syngenta had provided millions of dollars in funding, entomologist Ben Woodcock led a study that claimed neonicotinoid pesticides negatively impact honeybees. Bayer and Syngenta fought back against Woodcock's work. "From a personal perspective," Woodcock responded, "I don't really appreciate having them accuse me of being a liar. And accusing me of falsifying results by cherry-picking data."[42] But Bayer and Syngenta did not really appreciate the implications of those research results, and therefore challenged the researcher.

CLAIMS OF INCOMPETENCE

A challenge to scientific research can also be a challenge to the scientist's competence. Question the choice of subjects, the handling of the samples, and the statistical test. A toxicologist who worked for the Ethyl Corporation called an independent scientist's work on industrial lead contamination in the mid-1960s "remarkably naïve" and said he was "woefully ignorant" of the field.[43]

Accuse reporters of plagiarism or lack of expertise. Argue that the media are "gullible when it comes to swallowing the whole utterances of the doomsayers."[44] Claim that "a know-nothing from the *New York Times* believes that the pharmaceutical industry is intentionally hiding data from clinical trials."[45]

CLAIMS OF ULTERIOR MOTIVES

Call for a close inspection of the motives of scientists, reporters, and activists. Say they are doing their research for the money. In fact, they need a problem to exist so that they can get money. Remind the public that this is not about science or the truth, it is about fund raising. Scientists make grandiose and apocalyptic predictions to stay on the gravy train of federal funding.[46] Environmental groups are also motivated by money. "You know, they're hopeless fundamentally," University of Washington fisheries professor Dr. Ray Hilborn said about Greenpeace after they used a federal records request and revealed his millions of dollars in industry funding. Dr. Hilborn has denied any wrongdoing and the University of Washington found he had not breached their policies and procedures on conflicts of interest.[47] "They're basically a money-raising organization, and they have to scare people to raise money." Three CEOs of major tuna companies pointed out the same thing about Greenpeace in a *Wall Street Journal* opinion piece. "Unfortunately, this attack on canned tuna isn't about science. It's about fund raising, and Greenpeace has discovered a recipe for success: Target something that's easily recognizable (like tuna), make some scary claims in the media, parade around in funny costumes—and start raking in the donations."[48]

Accuse messengers of being motivated by status and jet-setting. Say they are just interested in the limelight. Accuse them of being publicity hounds (never mind that it is the Corporation that has internal and external PR experts, press releases, media consultants, external PR firms).

Call them elitists. A university expert who had also worked for a division of the American Cyanamid Company, a leading manufacturer of agricultural chemicals, in 1971 drew attention to the National Audubon Society, which was fighting against DDT, for their "predominantly white and middle-class membership."[49] He claimed the society "shows underlying resentment of human beings" and "has no program for the relief of suffering among millions of human beings in the tropics."[50]

AD HOMINEM ATTACKS

Personal attacks might be appropriate, per Hill & Knowlton's "smear and belittle" recommendation. Dr. Kevin Folta, who has received funding from Monsanto, has called reporter Carey Gillam a "hideous human" and "disgusting."[51] Biologist Tyrone Hayes told a reporter that Syngenta was behind "derogatory remarks about his appearance, his speaking style, and even his sexual proclivities."[52]

Amplify any unexpected disparaging news about scientists and scholars. The websites Planet of the Vapes and Vaping360 both posted stories about how professor Stanton Glantz, who researches tobacco and nicotine, was accused of sexual harassment by a former employee (Glantz and the university denied the allegations and settled out of court in 2018).[53] Expect all these efforts to result in a "chilling effect" that will help dissuade others from challenging the Corporation.

CASE: STALLING THE PRESERVATION OF ANTIBIOTICS FOR MEDICAL TREATMENT ACT

Scientific research has demonstrated a rise in "superbugs"—new strains of bacteria that cannot be effectively treated by any existing antibiotic drugs. The World Health Organization (WHO) called antimicrobial resistance, which in large part occurs in response to the overuse of antibiotics, "an increasingly serious threat to global public health that requires action across all government sectors and society." The animal agriculture industry uses approximately 80 percent of the antibiotics sold in the U.S., which are fed daily to healthy animals to improve growth rates and prevent infections in dense living conditions (a single shed for broiler chicken can contain 20,000 to 30,000 day-old chicks).[1]

Some European countries have implemented policies to restrict antibiotic use in animal agriculture. A federal bill first introduced in Congress in 2007 proposed to restrict the use of eight classes of antibiotics only to sick humans and sick animals. In June 2012, *Consumer Reports* published a "Meat on Drugs" report and launched a campaign to support these new federal regulations. In response, a coalition of organizations involved with meat and poultry production, led by the Animal Health Institute, which represents companies that develop and produce animal medicines, wrote a letter to Congress calling the report and campaign misleading. The industry argued that comparing animal and human data is invalid, voluntary measures by companies were a better route, and that "the issue of antibiotic resistance is scientifically complex and cannot be addressed with simple solutions." The bill was last introduced in 2017; it has never made it to a vote.[3]

What lessons can be learned from this approach to potentia

CHAPTER 8

CHALLENGE THE POLICY

If both the problem and its cause have been established in the public arena, then activists and regulators are likely to advocate for new rules. While the ultimate goal is to preempt and prevent legislation, if regulation is probable, there are still ways to stall. Just as creating scientific controversy can delay policy, a lively debate about the optimal policy approach can also buy time.

This section provides arguments to use against policy proposals. The Corporation can use its scientific experts to address policy, especially if the experts are comfortable speaking about issues outside their expertise (they should be by now). There are always economists for hire who can show how disastrously expensive policies will be, armed with tools like cost-benefit analysis that can be adjusted to provide almost any desired outcome.[1] It does not matter if these calculations fail to distinguish between "acts of omission and commission, proximate and non-proximate causes, negative and welfare rights, justified and unjustified paternalism, causal chains of responsibility," or anything else.[2]

Of course, some policies are good policies. A policy that provides the Corporation protection against certain lawsuits can be consistent with protecting the Corporation from the risks of scientific knowledge. Legally classifying civil disobedience by environmental and animal activists as terrorism can certainly help protect free enterprise. This kind of policy the Corporation might even help draft, and scale it with assistance from groups like the American Legislative Exchange Council. But, as with the previous sections about science, this section is focused on the threatening forms of policy, rather than beneficial forms.

MORE POLICY RESEARCH IS NEEDED

Policy can be delayed with drawn-out debates about which policy is best. Even after BP publicly accepted the existence of climate change, they remained "very conscious that there remains an enormous debate among the economists about the best way of addressing the issue."[3] More policy research, and therefore time, is needed. Put sand in the gears.

SUFFICIENT POLICY IS ALREADY IN PLACE

Insist that existing regulations are sufficient. Johns Manville held "the conviction that [they] were already doing everything possible to reduce risk. The asbestos company had modern dust-collection equipment and a standard for airborne fibers that bettered the national standard at the time by half. [They] also issued regular bulletins about acceptable procedures and exposure levels. What more could [they] possibly do?"[4] Syngenta has pointed out that new regulations for atrazine were not necessary because existing federal standards are already safe.[5] The fishing industry argues against marine protected areas on the grounds that current catch limits are enough.

THE POLICY IS TOO EXPENSIVE

Fund analyses to show that the policy is too expensive and would lead to disastrous financial consequences. Argue in vague terms about how harmful the policy will be. "What good is it to save the planet if humanity suffers?" asked ExxonMobil CEO Rex Tillerson.[6] Restrictions on fossil fuel use would lead to "rising consumer energy prices" that in turn "will depress the living standards of American families."[7] Climate policy will result in "higher energy costs [that] will handicap small business."[8] Smoke-free ordinances would hurt revenues in the restaurant business.[9]

Argue using precise terms, regardless of whether the numbers are backed up by analysis. An executive with the IPAA published a letter in *The Boston Globe* in which he cited an IPAA-commissioned report that concluded that divestment by the nation's top pension funds "could mean trillions in losses over the long term" and urged Massachusetts to "reject this empty gesture."[10]

THE POLICY IS A WASTE OF TAXPAYER MONEY

Almost every policy has some element that can be characterized as a misuse of taxpayer money or time, so invoke the taxpayer often. Monsanto insisted that government testing of food for glyphosate residues would represent a "misuse of valuable resources."[11] The trade association CropLife America claimed EPA meetings were a waste of taxpayers' time.[12]

THE POLICY WILL HURT WORKERS

The policy will be bad for workers, and lead to unemployment or unsafe working conditions. A representative from Hooker Chemical and Plastics Corporation claimed that a federal standard of "no detectable level" of vinyl chloride would mean plant closures and the loss of fifty thousand jobs.[13] Global warming policy would "devastate employment in major U.S. industries."[14] Marine protected areas and fisheries management take away freedom from fishermen, reduce their income, and endanger them because they will be forced to go farther offshore to fish. When the World Health Organization published a report on diet and exercise that suggested individuals "limit the intake of 'free' sugars," the Sugar Association threatened that they would try to block U.S. funding for WHO and insisted that their objections were on behalf of "the hard working sugar growers and their families."[15]

THE POLICY UNDERMINES CONSUMER FREEDOM OR INDIVIDUAL RIGHTS

Policies that restrict consumer or producer behavior will curtail consumer freedom. Restricting fossil fuel use means "that non-elected, unaccountable bureaucrats will gain greater control over our lives and resources."[16] A tax on soda would undermine the autonomy of the consumer. In 2011, the NRA said gun research would only advance "the false notion that legal gun ownership is a danger to the public health instead of an inalienable right."[17] Californians for Balanced Energy Solutions, a group established by the natural gas company SoCal Gas, cautioned on its website that there are "powerful organizations that are working to take away your right to choose affordable natural and renewable gas."[18]

THE POLICY HURTS POOR PEOPLE

The policy will hurt poor people. Use experts (someone with a PhD—any PhD will do) to make this argument. Someone with a PhD in animal science and funded by the beef industry can argue that "producing less meat and milk will only mean more hunger in poor countries."[19] A Monsanto-funded professor with a PhD in molecular biology can be paid to write briefs that "demonstrate how activists' messages and tactics regarding Genetically Modified (GM) crops and plant biotechnology undermine worldwide efforts to ensure a safe, nutritious, plentiful and affordable food supply" and "provide examples of activist campaigns that spread false information."[20] A professor with a PhD in zoology and funded by the fishing industry in developed countries can sanctimoniously advocate for the world's poor and claim that marine protected areas jeopardize the "world's poorest people who rely on marine fisheries for nutrition and income" and "closing the oceans would condemn them to starvation or abject misery."[21]

A soda tax is regressive and therefore disproportionately impacts poor people (so, too, does the sugar, but ignore that). Global warming policy will "kill the African dream" and condemn the poorest countries to "perpetual poverty"[22] (at least the ones who remain above sea level).

Regulations will hurt already disadvantaged children. A 1982 editorial in *The Wall Street Journal* noted that "studies have discovered high levels of lead in the bloodstream among certain children, particularly children already disadvantaged by life in urban ghettos. . . . On the other hand, it is seldom admitted in these debates that a more efficient economy, capable of producing more wealth and more jobs would most likely do more for ghetto children than some of the protections the environmental lobby is demanding."[23]

The poor can always be used strategically. The PR firm Burson-Marsteller helped the coal company Peabody design an advertising campaign to frame coal as a solution to poverty.[24] ExxonMobil, along with the rest of the fossil fuel industry, regularly reminds the public that its products are "essential for human society's progress" and enable "growth and prosperity."[25] Natural gas keeps energy prices low, which benefits low-income families. U.S. agribusiness works to "feed the world."[26]

THE POLICY COSTS LIVES

The policy might not just hurt people, it could kill them, or already has. Rachel Carson wrote *Silent Spring,* which generated hysteria that led to the banning of DDT, and as a result millions have died of malaria. In addition, many of the world's poor died of hunger because the benefits of pesticides were not realized (ignore the unsavory problem of unequal food distribution).[27]

THE POLICY WILL BE INEFFECTIVE

Insist that the policy will not work or will not solve the problem entirely. GM argued that it was unclear that seatbelts would work.[28] Divestment is merely symbolic and will not impact fossil fuel industry profits.[29] Even if marine-protected areas did work to protect biodiversity, they would be too difficult to monitor and enforce.

THERE ARE MORE IMPORTANT POLICIES

The policy is a distraction from more important policies. Campaigns to divest from fossil fuels divert our attention from better solutions.[30] "We should make sure that the focus on Marine Protected Areas does not divert attention from other strategies of ocean protection."[31] A spokesperson for the California Restaurant Association asked, "With crime and budget shortfall issues, why are city and state legislators focusing on trans fats and fast food restaurants?"[32]

THE POLICY IS ARBITRARY

The policy is arbitrary and therefore wrong. A DuPont official, in a 1976 letter to the president of the National Academy of Sciences, which was then studying the impact of CFCs on the ozone layer, argued that basing legislation on "unproven theory" would put the country on "the road to the rule of witchcraft where, by definition, the accusation proves the charge."[33] Hooker Chemical and Plastics Corporation objected to the proposed policy standard of "no detectable level" of vinyl chloride in the workplace because there was no evidence of angiosarcoma (cancer of the inner lining of blood vessels) for workers exposed to vinyl chloride at levels below 50 ppm.[34] Justin LeBlanc of the National Fisheries Institute called proposals to protect 20 percent of U.S. waters as no-fishing zones "arbitrary" and "completely irrational."[35]

THE POLICY WILL HAVE UNINTENDED CONSEQUENCES

Insist the policy will lead to negative and unexpected consequences in another domain. A policy to reduce greenhouse gas emissions would "hinder the advance of science and technology."[36] A policy to expand protected areas in the ocean will mean fewer fish caught, which could mean more beef production and more deforestation for crops or cattle grazing.[37]

THE POLICY THREATENS A COLLECTIVE IDENTITY OR NATIONAL SOVEREIGNTY

The policy undermines national identity or sovereignty. The Green New Deal would end the dairy business in Wisconsin, and "threaten the affordability and reliability of energy" and, in short, "destroy the Wisconsin way of life."[38] Climate policy that gives the "government control over energy production and consumption will strengthen repressive institutions, particularly in the Third World and the former Soviet Union," and "impair the readiness and training of the U.S. armed forces."[39]

THE POLICY IS UNNECESSARY BECAUSE THERE IS A TECHNOLOGICAL FIX

The policy is unnecessary because a technological solution is inevitable. These technological fixes, like geoengineering to address climate change, can be proposed alongside expressions of optimism and faith in human ingenuity. If there is not an actual technological solution, invent something that looks like one. By the mid-1950s, the cigarette companies each began manufacturing cigarettes with filters, which they claimed strained out tar and nicotine (they did not), to put consumers' minds at ease (which presumably they did).[40]

THE POLICY SHOULD NOT BE DETERMINED BY THIS GROUP OF PEOPLE

A policy can be argued to be illegitimate because of the identities of the people who champion or oppose it. Capitalize on popular social issues, which, at present, include issues of privilege and discrimination. "Meet the rich moms who want to ban vaping."[41] Individual companies and the American Beverage Association have called government proposals to regulate sugar-sweetened beverages "discriminatory" against soda.[42] After several banks, including Goldman Sachs, said they would not finance oil and gas drilling in the Arctic, a Native Alaskan mayor published an opinion piece in *The Wall Street Journal* titled, "Goldman Sachs to Native Alaskans: Drop Dead." Alaska's three members of Congress called on the federal government to "use every administrative and regulatory tool at your disposal to prevent America's financial institutions from discriminating against America's energy sector." They quoted from the mayor's opinion, and noted that the banks' refusals were not just discriminatory toward the energy sector but also toward Alaska Native communities (although some Native Alaskan groups supported the banks' stance).[43] That same summer, when the California town of San Luis Obispo wanted to introduce a proposed policy to reduce the use of natural gas, an energy industry media consultant pitched reporters that the policy was racially discriminatory.[44]

ANY POLICY IS OVERREACH

Equate any policy with the concept of government overreach. Regulation represents a step forward on the "slippery slope of government intrusion."[45] Claim the policy is illegitimate, even if it is nongovernmental policy. For instance, insist that universities should not divest from fossil fuels on the grounds that universities should be apolitical.[46]

CASE: HUMAN NATURE CAN HELP JUSTIFY INACTION

One popular explanation for inaction on climate change is that there is something fundamental to human behavior that has prevented us from acting. Jonathan Franzen, one of the few notable writers to regularly engage with environmental issues, acknowledged there was probably a period in the late 1980s when most of the negative effects of climate change could have been averted, but believes we were prevented from doing so by "the constraints of human nature." In other words, there is something fundamental to all humans (not just Americans, not just Republican members of Congress) that made climate change impossible to solve. "Finally, overwhelming numbers of human beings, including millions of government-hating Americans, need to accept high taxes and severe curtailment of their familiar life styles without revolting. They must accept the reality of climate change and have faith in the extreme measures taken to combat it," Franzen explained. "Call me a pessimist or call me a humanist, but I don't see human nature fundamentally changing anytime soon."[1] Ergo, inaction is justified as essential to human nature. The problem is human nature, and human nature is fixed.[2]

Are there genuinely independent and authoritative arguments that simply accidentally bolster or can be repurposed to defend free enterprise against scientific knowledge or government policies?

CHAPTER 9

OUTSIDE OPPORTUNITIES

Watch for outside opportunities—independent ideas and ways to characterize a problem that may assist the Corporation to sow doubt, redirect blame, or push the problem aside. Why buy these cows if the milk can be repackaged and sold for free? These truly independent arguments back up or assist the Corporation's arguments, and make challenges to science easier.

THE PROBLEM IS A RESULT OF HUMAN NATURE

The Corporation may want to amplify arguments that suggest biology, innate human tendencies, are the reason certain social problems, like dishonesty or climate change, exist. Duke University psychologist and behavioral economist Dan Ariely and his colleagues noted in one study that "fraudulent or dishonest actions are not exclusive to the realm of big corporations" but instead "dishonesty is part of the human condition."[1] We cannot fix a problem because we're not hardwired to fix the problem. (Never mind that we were also not hardwired to read—the visual stress of reading and computer work has exacerbated nearsightedness—or to scuba dive or to drink milk.)[2] "Our minds are not designed to respond to environmental problems when such problems are distant, global, and presented in abstract terms."[3]

The failure to adequately address climate change is due to something essential in human nature. Writer George Marshall published *Don't Even Think About It: Why Our Brains Are Wired to Ignore Climate Change.* Marshall quotes psychologist Dan Gilbert, who said that climate change is "a threat that our evolved brains are uniquely unsuited to do a damned thing about."[4] The argument that we are not designed to deal with climate change spans the full gamut of human psychology, from the innate tendency to discount the future to excessive optimism, our faith in the supernatural, and our attraction to technological fixes.[5] Arguments that shift blame from powerful producers to "human nature" or "all of us" are useful in distributing responsibility. Of course, the more that some sectors of society do to address climate change, the harder it is to argue inaction is a fundamental feature of human nature.

THE PROBLEM IS VERY COMPLEX

Another independent argument that can be used to the Corporation's advantage is that the problem is too difficult or impossible to fix. Climate change is often characterized by academic researchers as the world's largest collective action problem, as a "wicked problem" that can never be solved. Solving a problem is distinct from addressing it (most societies have addressed slavery, but have not ended it completely), but this framing may help demotivate society to attempt either one. There are also discussions about "competing knowledge claims" and "relativism" and the notion that "nothing can be truly answered by science." Recall that one of Hill & Knowlton's suggestions for dealing with tobacco experts was "to raise the issue far above them, so that they are hardly even mentioned; and then we can make our real case."[6] Arguments that things are complex help elevate the issue.

THE OPPORTUNITY TO ADDRESS THE PROBLEM HAS PASSED

Independent arguments that it is too late to solve a problem can also be useful. An article in *The New York Times Magazine*, "Losing Earth: The Decade We Almost Stopped Climate Change," ignores that there are degrees of disaster.[7] Whether it is climate change, nuclear waste, or income inequality—there may be opportunities to emphasize work that suggests there was a time when something could have been done, but that window of opportunity has closed.

THE NEXT GENERATION WILL ADDRESS THE PROBLEM

The Corporation can emphasize common notions that the new generation can address the problem, which serves to focus attention on the future rather than the present. A teenager becoming an icon of climate action is a good thing because it suggests those who want action are not in positions of authority. Leaders from the scientific community who give talks on "equipping our children to manage the planet" incidentally serve the goal of regulatory delay.

FOCUS ON HOUSEHOLD CONSUMERS

There is no supply without demand. Research that focuses on individual decision making at the household level is a helpful complement to active argumentation about consumer responsibility. The question is not what encourages a company to make a better product, but what makes a customer buy it. Research that conceives of individual consumers, not government oversight over large-scale producers, as responsible for how things are manufactured can be helpful.[8] Many academic fields, such as behavioral economics and psychology, focus specifically on the role of households and individual consumers and how to nudge them into better decision making rather than impose regulations.

Even the wide, independent literature on the psychology of climate denial deals almost exclusively with the audience response to messages and not at all with the psychology of those at the podium. (The famous Milgram shock experiments likewise examined the psychology of the people following orders, but did not address the psychology of the people giving them.) The field of psychology seems to show little interest in the decisions of Lee Raymond, the former CEO of ExxonMobil, who took the company's climate denial to its zenith, or any CEO for that matter, unless it is to highlight his exceptional talents. This is all to the Corporation's advantage.

LOOKING FOR VILLAINS IS UNPRODUCTIVE

The self-help and wellness factions of the independent research community largely agree that looking for villains is unproductive. Anger is unhealthy. A culture of naming, shaming, and blaming is not good for society. Dr. Adam Grant, an organizational psychologist at the University of Pennsylvania's Wharton School of Business, noted, "In the face of injustice, thinking about the perpetrator fuels anger and aggression. Shifting your attention to the victim makes you more empathetic, increasing the chances you'll channel your anger in a constructive direction. Instead of trying to punish people who caused harm, you'll be more likely to help the people who are harmed."[9] Do not seek retribution. Focus on self-healing. These are perfectly complementary messages to the strategy outlined in *The Playbook*.

CRITICISM OF UNIVERSITIES AND PROFESSORS

Independent efforts that denigrate universities and scholarship, especially the social sciences and humanities, are also helpful. Amplify arguments that political conservatives are underrepresented in particular fields or universities as a whole. Consider underwriting groups like Campus Reform—a "conservative watchdog" for higher education that "exposes liberal bias and abuse on the nation's college campuses."

Look for ways to piggyback on moments when university experts, especially those in the natural sciences, are critical of their colleagues who engage in activism of any kind. Harvard University professor and historian Allan Brandt (author of the 2007 book *The Cigarette Century*) said he testified only once against the tobacco industry because, according to Brandt, "it's enormously time-consuming and labor-intensive to testify." In his book he wrote that he had "no interest in becoming an expert witness. . . . I did not want my scholarship to be dismissed as 'advocacy.' "[10] That swaths of university experts believe that being objective means not engaging with anything political or legal has helped neuter an entire class of intelligent and independent professionals. This works to the advantage of the Corporation.

University experts, especially scientists, often have trouble clearly conveying their results, and these poor communication skills as well as a reputation for poor communication skills help the Corporation. The PR firm Hill & Knowlton noted in a memo to their tobacco manufacturers that the on-air performance of Alton Ochsner, a surgeon and antitobacco expert, "was so terrible that it would do us more good than harm."[11]

Anyone blaming poor science communication skills for the failure to create a successful mass movement is also useful. "Clearly there is something in how we communicate climate change that is failing to mobilize a wider audience," wrote two experts of climate change communications.[12] The executive director of the United Nations Environment Programme pointed out that the "language of environmentalists has been boring, so uninspiring . . . with many acronyms and politically-correct phrases, no one will listen." He noted, "You cannot bore people into action. They need to be excited and inspired to take action and change their behaviour."[13]

The more that scientists are willing to take the rap for policy inaction, the better. One atmospheric physicist told a reporter that "climate change has become a religion," which is part of the reason for the polarization because "we just ask people to believe it. And we're not willing to engage with the skeptics."[14] Another professor put it, "We'll never tackle climate change if academics keep the focus on consensus."[15] Professors are the problem.

SOCIAL CHANGE HAPPENS SLOWLY

Given the Corporation's goal is to postpone action, amplify the voices of those who argue that social change happens slowly. It is encouraging when scholars argue that "deep reform of the corporate structure, if it happens, will likely unfold slowly."[16] The Corporation is served well by the argument that social change is slow because it justifies regulatory delay (even if every company privately knows that it is possible to get people to do almost anything quickly).

TECHNOLOGICAL SOLUTIONS

In every problem, there is a market opportunity. Problems are to be solved, not prevented. Technological solutions are preferred to any bans on limits of production. Rather than focusing on leaving oil in the ground, focus on the opportunities to profit from adapting wealthy coastal communities for climate change. Patents can be filed for geoengineering technologies to sequester carbon. Cows can be fed algae to reduce their methane emissions. Species can be saved with cloning and gene editing technology, which can raise venture capital, unlike other interventions like habitat preservation and reforestation. Seize on the widespread belief that the class of people most capable of technological innovation are capitalists and entrepreneurs[17] because, after all, this means a clear role for (and an absolution of) the Corporation.

CASE: GROWING DISTRUST IN BIG BUSINESS

The heyday of high confidence in the Corporation may be waning. The 2020 Edelman Trust Barometer, which surveyed 34,000 people across twenty-eight different markets on their trust in business, government, NGOs, and media, reported that business is seen as competent but unethical. More than half of the population surveyed agreed with the statement: "Capitalism as it exists today does more harm than good in the world," and 76 percent worry about fake news being used as a weapon.[1] The U.K. expressed its lowest trust ever in all aspects of life, and, as a country, only Russia expressed more distrust.[2]

A survey of 4,039 U.S. adults showed that Americans' trust in big business declined sharply between 2018 and 2019, flipping from a majority positive view of big business to a majority negative one.[3] Another recent survey found that 58 percent of U.S. adults reported trusting scientific research findings less if they learn that they were financed by industry.[4] The same survey found that U.S. adults trust research funded by the government twice as much as research funded by industry.

How can industry funding for scientific knowledge be hidden to avoid the distrust associated with it? How can businesses ensure a continued social license to operate?

CHAPTER 10

NEAR-TERM THREATS

The Corporation has done very well in the twenty-first century in terms of sheer financial and political might, as well as positioning itself culturally as an institution that has concerns beyond shareholder value. The Corporation has also successfully used *The Playbook* to achieve widespread success in delaying regulations. But a rising skepticism about the Corporation and calls for accountability are now coming from multiple directions. Some near-term threats to business operations related to the Corporation's treatment of scientific knowledge are detailed here.

INTERNAL STRIFE AND TALENT RETENTION

Recent cases of internal strife and failure to retain talent over scientific denial and policy inaction are cause for concern. It is one thing when a reporter uncovers denial by the asbestos industry, and another thing when a former manager at an asbestos company reveals how that company denied the harms of asbestos in the pages of the *Harvard Business Review*.[1] It is likewise one thing when a handful of professors or activists get worked into a lather over climate denial, and another when a handful of top executives publicly leave a top PR firm because it has accounts from climate-denying organizations, as Edelman executives did in 2015.[2] Hundreds of employees at Amazon in 2020 criticized the company in an open letter for failing to act on climate change.[3] A recent survey on workforce sustainability found that "office workers are not only more likely to leave companies that don't implement sustainable business practices, they will also speak out on public forums—turning internal issues into external reputational damage."[4]

If people once close to a company, an industry, or a product turn against the product they once helped make, the discord that results can be difficult to explain. Wendell Potter, a former health insurance executive, turned against the insurance industry. The Rockefeller Brothers Fund—a large charitable trust made possible by the wealth generated by Standard Oil—announced its divestment from fossil fuels, the very commodity that had created its endowment. These cases are rare, but can get a lot of attention, and the Corporation should aim to prevent them. In addition, growing legal protections for whistleblowers make internal dissent even more likely. It is rumored that an executive for a large meat and dairy company will soon come forward with evidence about how their company denied scientific findings related to public health, animal welfare, and climate change.

STUDENT ACTIVISM OPPOSING INDUSTRY FUNDING TO UNIVERSITIES

Student campaigns aimed at ending industry relationships, such as the UnKoch My Campus campaign (to stop Charles Koch's "takeover" of higher education), are soon to spread internationally. A campaign is in the pipeline for which students will demand that the universities they attend "put knowledge first." The student groups are claiming that industry donors are exerting undue influence on research, which is jeopardizing their future. The students have pressured universities to refuse "dirty" research money and funding agreements that come with any restrictions on academic freedom. After long debates with students, it seems both Oxford University and Harvard University will soon announce funding and full access to university archives for two doctoral students to write histories of industry-funded scientific research at each institution.

UNIVERSITY DISCLOSURE POLICIES

The current momentum is toward more disclosure in scientific research, not less, and more research into industry funding's impacts on research results. A collaboration between *ProPublica, The Chronicle of Higher Education,* Greenpeace, and the Union of Concerned Scientists resulted in a searchable online database published in 2019 of the absolute and relative amounts of corporate money entering various universities from the private sector, including grants and consultancies to faculty.[5] Databases of scientific articles are now adding acknowledgments and competing interest statements to searchable bibliographic records. New software is making it easier to study patterns of funding. One recent study located 389 articles from 169 different journals that included 907 researchers who acknowledged funding from Coca-Cola. Researchers cross-checked that list of 907 authors with a list of experts to whom Coca-Cola acknowledged giving money, and they could find just 42 (less than 5 percent), leading the study's authors to conclude that "the Coca-Cola Company appears to have failed to declare a comprehensive list of its research activities."[6]

It is rumored that some universities, hoping to restore their reputations with their students and the public, will introduce material consequences for faculty who fail to disclose funding. According to sources, some university faculty senates have discussed an open pledge to never take money from any party that demands to see the results before publishing. There are university units, especially in the medical division, attempting to hold experts accountable for their public and political engagements, and some are proposing that all faculty provide an "electronic long-form COI [conflict of interest] disclosure statement" that is available online. Experts are expected to "list all financial/funding relationships, including honoraria and travel reimbursements or other compensation."[7]

MEDIA POLICIES

There are discussions among major news outlets, including *The New York Times,* to set a standard disclosure policy for opinion pages and to require reporters to include details about industry funding for quoted sources. Some newspapers now have rules about advertising that promotes falsehoods. Fortunately, many, like *The Wall Street Journal,* continue to accept "a wide range of advertisements, including those with provocative viewpoints."[8]

Some newspapers are considering requiring that their reporters include information on funding when they write about scientific research. Up until now, even when industry funding is declared in the press release or in the scientific article, most reporters would not include these details in their stories. A UC Davis press release about research that challenged a U.N. study linking meat production to climate change noted that the study had been funded by the beef industry, but major news outlets across the U.S., Canada, and France that reported on these contradictory findings never mentioned the research's industry funding or the lead author's ties to the beef, pork, and dairy industries.[9] It's not just cows. When a science reporter for *The Atlantic* covered a conflict between one scientific study that found more than half of the ocean was industrially fished, and another that found that industrial fishing occurred in just 4 percent of the ocean, he left out that the latter study was funded by sixteen different seafood industry groups.[10] These new policies will require reporters to cite funding for studies and experts they quote, and prevent this from happening in the future. These new requirements would create some additional hurdles for the Corporation.

DISINFORMATION POLICIES

Disinformation is now high on many agendas, especially with the alleged role it has played in democratic elections. Authentication services are expected to rise. In the U.S., Senator Sheldon Whitehouse (D-RI), who has already spearheaded an effort to combat climate disinformation, is rumored to be forming a political coalition that will undertake a sweeping policy effort to tamp down on disinformation more broadly, including holding social media platforms accountable as publishers. In 2020, several global social media platforms instigated their boldest internal policies to date to combat disinformation and "hate speech," but more policies are rumored to be in the pipeline. The BBC's anti-disinformation unit recently received a new influx of funding.[11]

There is also a team of international leaders who may soon propose a Framework Convention on Disinformation Control as a possible legally binding treaty. There are also rumors of adding an eighteenth goal to the U.N.'s Sustainable Development Goals on "evidence" that would commit countries to the right of all people to the highest and best standard of scientific evidence on all issues that affect sustainability. Any attempts to seriously address disinformation are likely to significantly impact the Corporation's scientific denial campaigns.

LAWSUITS

The countless potential lawsuits related to the cover-up or subversion of scientific knowledge deserve their own memo. In brief, one outcome of having successfully delayed regulations could be decades of costly litigation. Recall that Johns Manville was not bankrupted by regulation, but by the 16,000-plus asbestos worker lawsuits in the early 1980s.[12] Fossil fuel companies are the defendants in a rash of lawsuits around the world that are challenging many legal traditions and affecting perceptions of company value.

Lawsuits are time-consuming, expensive, and have uncertain outcomes, but even if a case is dismissed, the pretrial discovery can do damage by uncovering incriminating documents. Some argue that the discovery and public attention of ExxonMobil's record on climate change is partly why Exxon says publicly that they support a carbon tax (meanwhile, privately, they lobby against climate policies and support lawmakers who do the same).[13] The number of databases that provide a repository for once secret company documents is growing and large repositories now exist for the tobacco industry, chemical manufacturers, and the fossil fuel industry. The more evidence there is that the Corporation denied scientific evidence and stifled policy efforts, the more public opinion seems swayed in favor of regulation and retributive justice.

UNIVERSITY PROGRAMS AND PROJECTS

There are faculty at several universities who are insisting on "ethically credible partnerships" as a standard, which means refusing agreements in which study design is in any way influenced by the funder, refusing agreements that condition publication on certain kinds of results, widely publicizing the partnership agreement and collaborative opportunities to the public and employees, and ensuring both partners are aware of other funding relationships each may be involved in.[14] There are also proposals that think tanks, especially those housed within universities, must be held to similar academic standards.[15] The Corporation should immediately begin planning workarounds for these imminent and onerous rules.

In addition, because the outcomes of lawsuits rely heavily on judges and future judges (who are currently law students), another scheme that poses a significant threat is the forthcoming Law and Knowledge Program, rumored to be supported by a cadre of wealthy philanthropists committed to "democratic values and scientific integrity." The program is supposedly being modeled off the highly successful Law and Economics Program that nearly half of federal judges attended between 1976 and 1999, which had support from at least 105 corporations.[16] After attending Law and Economics training, judges who participated used more economics language, and issued more conservative verdicts in economics cases, and more often ruled against regulatory taxation agencies.[17] The Law and Knowledge Program is applying this blueprint to disinformation with the goal of better scientific understanding and harsher penalties for disinformation and deception. Like the Law and Economics Program, the Law and Knowledge Program will include a curricular program at top law schools, as well as intense trainings at attractive resorts for both state and federal judges. This is expected to work against the Corporation.

The University of Washington is planning to add a Disinformation Department to its Information School. The Moore-Sloan Data Science Environments appears to have plans to make a large investment to study online disinformation. Teams of data scientists will work to uncover pushback and manipulation of scientific knowledge in real time, similar to how a recent study examined efforts to discredit a particular scientific effort on social media.[18]

Stanford University supposedly has plans to construct a new Museum of Agnotology devoted to the study of the creation of ignorance. It is rumored that the museum's mission will be to use exhibitions and educational programs to encourage inquiry into the structures and processes that impede the production of knowledge. An exhibit on "The Exxon Files" is proposed for the new Rachel Carson wing and will display materials from decades of scientific denial (probably to placate students, who have protested the large amount of Exxon money the university has taken). The museum will supposedly offer journalism fellowships and semester-long undergraduate courses. As Stanford has yet to break ground, there may still be time to obstruct this project.

Finally, a more immediate project of concern is one from a professor at New York University. An email about her new manuscript, which apparently describes much of the material in this *Playbook,* is attached here as an appendix. Any attempts to discredit her would likely only bring more attention to the project. The recommended course of action is to ignore her.

APPENDIX

From: Jennifer Jacquet <jj@jenniferjacquet.com>

Subject: The Playbook

Date: October 20, 2020, at 8:55:55 PM PDT

To: Dan Frank <dantheon@penguinrandomhouse.com>

Dear Dan,

As you know, this is not the subject I originally proposed. But the politics of 2016–2020, including a resurgence of climate denial, led me to want to better understand the challenge to science that originates with a company or an industry, motivated by money and market access. We discussed the new direction over lunch. You were encouraging.

I wanted to focus on the similarities between and across scientific denial that share a corporate origin. From the manufacturers of radium, asbestos, lead, tobacco, pharmaceuticals, firearms, sugar-sweetened beverages, fossil fuels, and meat, scientific denial has become a routine part of business operations: a private sector strategy that prioritizes money over truth.

Science can be so powerful that the powerful want to control it. The outlines of the strategy to challenge science can be elusive and it can take years or decades to even partially make sense of, in no small part due to secrecy of the corporation and its network of accomplices. But after a century of scheming, during which the tactics have been refined, disseminated, scaled, and globalized by public relations firms, it is clear that corporate scientific denial also has a particular gestalt, with a four-pronged pattern to the approach and the arguments: challenge the problem, challenge causation, challenge the messenger, and challenge the policy.

Each maneuver has some sleight of hand, but the end result of this strategy is not merely a card trick. Everyone appreciates that a card trick is in fact a trick. The end result is more like the casino, with its calculated architecture and design—the dimmed chandeliers, the comfortable furniture, the dealers, the drinks—to keep the people inside comfortable and gambling as long as possible. It is clear that the odds favor the House, but somehow the overall design distracts us from that fact. As with a casino, the strategies and success of corporate denial are at once hidden and obvious.

We have each experienced these forces in our society, and many researchers have felt the not-so-invisible hand of the private sector. Even I have had encounters, however minor they are in comparison to others', that are reminders of how prevalent this strategy is. In 2010, moments after writing a blogpost (the chosen medium of the powerful) admiring Jane Mayer's first article about the Koch brothers, I received notification that @KochFacts was following me on Twitter, and Koch Industries contacted the editors to dispute the post. The publication of my research has been delayed and arbitrated by lawyers for scientific journals, who are worried about lawsuits from industry. I have watched corporate intimidation deter my colleagues from asking certain kinds of questions. I have seen the industry co-opt the notion of conflict of interest and change the meaning to something other than financial interests, sometimes in opposition to my own research. I have conducted research on fisheries, climate change, and animal agriculture and encountered firsthand the industry shills, the power of industry funding, and the threat of litigation.

The tactics discussed here are the ones focused primarily on the institution of science, and part of why these tactics have been so successful is that they focus on a form of inquiry (science) and a group of people (scientists) unaccustomed to thinking politically. The culture of science is one in which people are trained and encouraged to think, and certainly to write, apolitically. These conditions explain why scientists are sensitive to accusations of bias or oversimplification or seeking the spotlight. The epistemology of science, with its norms of deliberate and dispassionate presentation of evidence, discourages those things. Scientists are likely to respond to such accusations in an earnest way—to accept an invitation to a climate science debate, to take personally accusations of being a "publicity hound," to make full-throated attempts at better science communication (believing the public has a gap in knowledge, rather than understanding that the fundamental gap is related to power), and to assume industry and its network are likewise interested in the truth, when in fact they are interested in regulatory delay or simply serving their clients.

I have listened as independent scientists think aloud about funding for research and come to the conclusion that being wined and dined by industry is no different from being wined and dined by anyone else. They naively believe that all money is money. Substantial evidence shows this is not the case. Money is money, but motives can lead to important differences in how that money is used to execute research. Industry funding is predicated on results that ultimately favor the industry's financial interests. Because of that, industry funding often comes with a unique set of conditionals, including agreements to suppress unfavorable findings or withhold final payments until both parties agree on the results. All money has some strings attached, some strings are more like leashes, and some come with collars. Most researchers are not aware of these differences, and that serves corporate interests.

In the same way that a casino can affect the character of a town, corporate-funded scientific denial has contributed to the erosion of scientific authority and mistrust in the government. In this casino, however, we are gambling with our health, the planet, and our most reliable way of knowing the world. The stakes could not get higher.

I hope that scientific institutions begin to recognize the dangers of industry-funded research and put in place safeguards to show they value truth more than money. A fortress must be erected that can protect scientific knowledge. Right now, that task is falling to individuals. Many researchers have refused to sell out to industry. Publishers have refused to pander to the omnipotent. Journalists have not been scared off their beats. Writing this has shown me a valiant set of people who value truth more than money. Thank you, Dan, for being one of them.

I must also thank Maria Goldverg at Pantheon, and Helen Conford and Casiana Ionita at Penguin UK, and the entire team at Brockman, Inc. I am also grateful to Nick Lepard, Kate Barrett, Genevra MacPhail, Dale Jamieson, Chris Schlottmann, and the twenty NYU students who took my course on this topic in 2019.

New York comes up so often in these corporate denial cases that I took the students in my class on a tour of denial through the city. But just as there is a geopolitics to ignorance—businesses in small towns have long been able to suppress unwanted stories in local newspapers as the town's main employer—there is also a geopolitics to knowledge. Scientific denial has been both created but also uncovered in New York, a city as crowded with hustle and greed as it is with fearlessness and decency. New York is headquarters to the deep pockets of denial, as well as intrepid individuals and institutions relentlessly trying to shake loose the tight grip on what the powerful permit us to know.

Eventually, we met for another lunch: me agitated over the state of reading, you your usual cool. I asked whether the twenty-first century really needed another *book*? You said it could be short.

Attached please find my attempt at epistolary nonfiction, except that corporate consultants write reports, not letters. They use the passive voice, bad figures, and enumerated lists. In an effort to prevent a total bludgeoning by this style, I took a different tack in chapter three, but the second-person form I chose may be equally irritating. Looking forward to discussing the format, visuals, and potential lawsuits (thank you to the lawyers who vetted everything).

I hope that you have been seeing a lot of your grandson. Zelda has been a great distraction.

As always,

JJ

P.S. You were right. What I sent previously was "not an ideal concluding chapter."

GLOSSARY OF TERMS

activist (also: *crusader, evangelical, fanatic, ideologue, zealot):* a term that can be applied to scientists and reporters to suggest they are ideologically motivated

advertorial (also: *op-ad*): a paid advertisement that allows the Corporation to present its position on an issue

agenda: a term to describe the efforts of scientists, activists, and reporters

alarmist (also: *doom and gloomer, pessimist*): a term to use against anyone who delivers an unwelcome scientific message and panders to a media landscape that rewards extreme messages

alternative hypothesis: another explanation for a particular result

animal experiments: depending on how the results of the study are interpreted, these are either the highest scientific standard or irrelevant (also see: *standards of evidence*)

balance (also: *both sides*): when the media include competing viewpoints on scientific issues

biased (also: *one-sided*): not neutral, in a way that is unfair

boring (also: *nothing new, the same old, yawn*): a term to use to disparage both individuals and scientific results

chilling effect: successfully deterring individuals from pursuing particular kinds of research, from speaking to reporters, or from giving courtroom testimony

common sense (also: *intuition*): appeal to gut reaction rather than conscious reasoning

conflicts of interest (COI) (also: *competing interests*): focus on religious beliefs, volunteering, dietary preferences, and political memberships—anything to conflate or distract from the original meaning of the term, which was that financial interests may compromise research

conspiracy: use to imply that unfavorable scientific results are part of a secret plot (also see: *hoax, myth, propaganda*)

consultants: experts for hire, especially to defend against scientific results (also see: *experts, think tank, university experts*)

consumer advocacy groups (also: *grassroots*): create these to fabricate opposition to scientific findings or policy (also see: *third-party allies*)

controversy (also: *no real proof*): a term to challenge the validity and meaning of scientific findings

correlation: depending on how the results of the study are interpreted, either an insight into causation or irrelevant

denial of scientific knowledge: the challenge to a certain kind of information with the goal of regulatory delay; a fiduciary duty of any company

detection bias: finding an increase in a phenomenon due to looking for it in a new or different way

discovery: formal process of exchanging evidence that may be presented at a trial, and which may be made public during or after litigation

echo chamber: when arguments are insulated from rebuttal

emotional: disparaging term to use against scientists, reporters, and activists, especially women

euphemism: a more innocuous term; e.g., “climate change” instead of “global warming”

experts: voices of scientific authority (also see: *consultants, think tank, university experts*)

fiduciary duty: the moral obligation to increase company value for shareholders

FOIA (Freedom of Information Act): a federal law that allows for soliciting documents, including emails, from university experts at state universities

good epidemiology: a concept to demand certain scientific standards in the study of diseases (also see: *sound science*)

government overreach: how to characterize any policy

hoax: a term to denigrate a body of scientific evidence (also see: *conspiracy, myth, propaganda*)

holding strategy: delaying regulation, the primary goal of denying scientific knowledge (also see: *regulatory delay*)

junk science: a term to denigrate unfavorable scientific findings (antonym: *sound science*)

misleading: term to denigrate scientific studies

model: depending on how the results of the study are interpreted, it is either highly flawed or the best all-purpose tool (also see: *standards of evidence*)

myth: a term to denigrate a body of scientific evidence (also see: *conspiracy, hoax, propaganda*)

natural variability: the explanation for a particular result that attributes causation to something in nature (rather than humans)

personal responsibility: the cause of any problem can ultimately be attributed to individual consumers, rather than institutional producers

press release: an essential device to frame arguments and influence public opinion

propaganda: a term to denigrate a body of scientific evidence (also see: *conspiracy, hoax, myth)*

public relations: the private sector equivalent to government propaganda

publicity hounds: term to disparage scientists who speak to the media that implies they are seeking attention for a cause or themselves

reanalysis: demand that a new scientist or group of scientists offer an outside assessment of the evidence

regulatory delay: delaying regulation, the primary goal of denying scientific knowledge (also see: *holding strategy*)

retraction: demand the withdrawal of a scientific study

sample size: depending on the results of the study either sufficient or inadequate or not representative

scientific knowledge: the most reliable form of knowledge in the history of human civilization

self-regulation: preferred to any other form of regulation

slippery slope: any policy leads to government overreach

sound science: a concept to demand certain standards of evidence (antonym: *junk science;* also see: *good epidemiology*)

standards of evidence: any aspect of scientific inquiry that can be challenged as inadequate (also see: *animal experiments, model, statistics, uncertainty*)

standby press release: a device to frame arguments and influence public opinion on hand in anticipation of a crisis event

statistics: depending on how the results of the study are interpreted, either the highest scientific standard or inappropriate (also see: *standards of evidence*)

think tank: experts for hire (also see: *consultants, experts, university experts*)

third-party allies: individuals or institutions who appear to be independent supporters (also see: *consumer advocacy groups*)

uncertainty: depending on how the results of the study are interpreted, either acceptable or unacceptable level of validity (also see: *standards of evidence*)

university experts: the most trusted scientific authority by the public (also see: *consultants, experts, think tank*)

victim: a stance the company or product can take if faced with extremely threatening scientific knowledge or policy

whistleblowers: employees who make risky information public

NOTES

1. DENIAL: A FIDUCIARY DUTY

1. Harvard Business School's Michael Porter told Joel Bakan that corporations are "the most powerful force for addressing the pressing issues we face." Joel Bakan, *The Corporation: The Pathological Pursuit of Profit and Power* (Toronto: Viking, 2004), 26.
2. Bakan, *The Corporation,* 26.
3. Lynn Stout, "Corporations Don't Have to Maximize Profits," *New York Times,* April 16, 2015.
4. Madeleine I. G. Daepp et al., "The Mortality of Companies," *Journal of the Royal Society Interface* 12 (2015): 20150120.
5. Michael Strevens, *The Knowledge Machine: How Irrationality Created Modern Science* (New York: Liveright, 2020), 32.
6. Steven Shapin, *The Scientific Revolution* (Chicago: University of Chicago Press, 1996), 165.
7. Dale Jamieson, personal communication.
8. Robert N. Proctor, *Golden Holocaust: Origins of the Cigarette Catastrophe and the Case for Abolition* (Berkeley: University of California Press, 2012).
9. Inmaculada de Melo-Martín and Kristen Intemann, *The Fight Against Doubt: How to Bridge the Gap Between Scientists and the Public* (New York: Oxford University Press, 2018).
10. Stephan Lewandowsky, Klaus Oberauer, and Gilles E. Gignac, "NASA Faked the Moon Landing—Therefore (Climate) Science Is a Hoax: Anatomy of the Motivated Rejection of Science," *Psychological Science* 24 (2013): 622–33.
11. Fred Panzer, "The Roper Proposal," email to Horace R. Kornegay, May 1, 1972, https://www.industrydocuments.ucsf.edu/docs/ltfn0108.

12. Linsey McGoey, "The Logic of Strategic Ignorance," *British Journal of Sociology* 63 (2012): 553–76.
13. Gerald Markowitz and David Rosner, *Deceit and Denial: The Deadly Politics of Industrial Pollution* (Berkeley: University of California Press, 2002).
14. Harry Frankfurt, *On Bullshit* (Princeton: Princeton University Press, 2005).
15. "Preliminary Recommendations for Cigarette Manufacturers," December 24, 1953, American Tobacco Records, Master Settlement Agreement, https://www.industrydocuments.ucsf.edu/docs/rsfy0137.
16. Thomas Whiteside, "A Cloud of Smoke," *New Yorker,* November 30, 1963.
17. Allan M. Brandt, "Inventing Conflicts of Interest: A History of Tobacco Industry Tactics," *American Journal of Public Health* 102 (2012): 63–71.
18. "Preliminary Recommendations for Cigarette Manufacturers," December 24, 1953, American Tobacco Records, Master Settlement Agreement.
19. Steven Koonin, "Can We Ever Get to a Zero-Emission World?," talk at iSEE Congress 2016.
20. Lee Fang, "The Playbook for Poisoning the Earth," *Intercept,* January 18, 2020.
21. https://www.syngenta.com/company/media/syngenta-news/year/2020/2019-full-year-results.
22. Steve Jarvis, "Narrative Talc—NTP Regulatory Challenge," http://asbestosandtalc.com.
23. Robert J. Brulle et al., "Obstructing Action: Foundation Funding and US Climate Change Counter-Movement Organizations," *Climatic Change* 166 (2021): 17.
24. Kevin C. Elliott, "Addressing Industry-Funded Research with Criteria for Objectivity," *Philosophy of Science* 85 (2018): 857–68.
25. Mayanna Lahsen, "Buffers Against Inconvenient Knowledge: Brazilian Newspaper Representations of the Climate-Meat Link," *Desenvolvimento e Meio Ambiente* 40 (2017): 17–35.
26. David Michaels and Celeste Monforton, "Manufacturing Uncertainty: Contested Science and the Protection of the Public's Health and Environment," *American Journal of Public Health* 95, Suppl. 1 (2005): S39–48.
27. Felix Wormser, Annual Meeting of the Members of the Lead Industries Association 4 (Lead Industries Association, 1935), cited in

Jenny White and Lisa A. Bero, "Corporate Manipulation of Research: Strategies Are Similar Across Five Industries," *Stanford Law & Policy Review* 21 (2010): 105–34.

28. Paul Brodeur, "The Asbestos Industry on Trial," *New Yorker*, June 17, 1985, 45.
29. Fred Panzer, "The Roper Proposal," May 1, 1972, Philip Morris Records, Master Settlement Agreement, https://www.industrydocuments.ucsf.edu/docs/ltfn0108.
30. Craig L. Fuller and Kathleen Linehan, Presentation to the Board of Directors, Philip Morris, June 24, 1992, http://legacy.library.ucsf.edu/tid/kgr52e00.
31. Whiteside, "A Cloud of Smoke."
32. Michael Specter, *Denialism: How Irrational Thinking Harms the Planet and Threatens Our Lives* (New York: Penguin, 2009).
33. Mark Wilson, "The New England Journal of Medicine: Commercial Conflict of Interest and Revisiting the Vioxx Scandal," *Indian Journal of Medical Ethics* 1, no. 3 (NS) (July–September 2016): 167–71.
34. Proctor, *Golden Holocaust.*
35. John Horgan, "What Thomas Kuhn Really Thought About Scientific 'Truth,'" *Scientific American*, May 23, 2012.
36. Bill Sells, "What Asbestos Taught Me About Managing Risk," *Harvard Business Review*, March–April 1994.
37. Alain Cohn, Ernst Fehr, and Michel André Maréchal, "Business Culture and Dishonesty in the Banking Industry," *Nature* 516 (2014): 86–89.
38. Jennifer Hiller, "New York City Sues Exxon, BP, Shell in State Court over Climate Change," Reuters, April 22, 2021.
39. Maxine Joselow, "Exclusive: GM, Ford Knew About Climate Change 50 Years Ago," *E&E News*, October 26, 2020.
40. Marc S. Reisch, "DuPont, Chemours Settle PFOA Suits," *Chemical and Engineering News*, February 15, 2017.
41. Carey Gillam, *Whitewash: The Story of a Weed Killer, Cancer, and the Corruption of Science* (Washington, DC: Island Press, 2017).
42. Peter Galison and Robert Proctor, "Agnotology in Action: A Dialogue," Chapter 2 in *Science and the Production of Ignorance: When the Quest for Knowledge Is Thwarted*, edited by Janet Kourany and Martin Carrier (Cambridge: MIT Press, 2020), 33.
43. Joselow, "Exclusive: GM, Ford Knew About Climate Change 50 Years Ago."

44. Philip J. Hilts, "Cigarette Makers Debated the Risks They Denied," *New York Times,* June 16, 1994.
45. Matthew L. Wald, "Pro-Coal Ad Campaign Disputes Warming Idea," *New York Times,* July 8, 1991.
46. Robert N. Proctor, "'Everyone Knew but No One Had Proof': Tobacco Industry Use of Medical History Expertise in US Courts, 1990–2002," *Tobacco Control* 15, Suppl. 4 (2006): iv117—iv125.
47. Proctor, *Golden Holocaust.*
48. John Bach, "Anti-Smoking Crusader," *UC Magazine,* May 2009.
49. Lawrence Carter and Maeve McClenaghan, "Exposed: Academics-for-Hire Agree Not to Disclose Fossil Fuel Funding," *GreenPeace Energy-Desk,* December 8, 2015.
50. Marc Parry, "Princeton Climate Skeptics Tried to Ignore a Campus Skeptic. Then He Went to the White House," *Chronicle of Higher Education,* August 16, 2019.
51. Núria Almiron, Natalia Khozyainova, and Lluís Freixes, "'This Nagging Worry About the Carbon Dioxide Issue': Nuclear Denial and the Nuclear Renaissance Campaign," Chapter 11 in *Climate Change Denial and Public Relations: Strategic Communication and Interest Groups in Climate Inaction,* edited by Núria Almiron and Jordi Xifra (Oxfordshire: Routledge, 2019).
52. Mike S. Schäfer, "Climate Change Communication in Germany," *Oxford Research Encyclopedia of Climate Science,* September 2016.
53. Geoff Dembicki, "How Exxon Silences Staff Alarmed by the Climate Crisis, According to a Former Employee," *Vice,* October 29, 2020.
54. Marc Gunter, "Edelman Loses Executives and Clients over Climate Change Stance," *Guardian,* July 7, 2015.
55. Brodeur, "The Asbestos Industry on Trial," 62.

CASE: DIVERSE ALLIES IN THE ENERGY SECTOR

1. Shell Oil, "Engineering Real-Life Heroes with Letitia Wright," YouTube, June 25, 2018.
2. Rebecca Leber, "These Ladies Love Natural Gas! Too Bad They Aren't Real," *Mother Jones,* December 14, 2020.
3. Western States and Tribal Nations Natural Gas Initiative (WSTN), San Juan College School of Energy, Sempra LNG, Navajo O&G, Black-Hawk Energy Corp., Consumer Energy Education Foundation, ETAC

RFI SUBMISSION FORM, October 30, 2020, https://tinyurl.com/jbmbmha8.

4. Western States and Tribal Nations Rural Gas Initiative (WSTN), Form 990-EZ, 2019, https://tinyurl.com/2msd38v9.
5. Application of Southern California Gas Company (U 904 G) and San Diego Gas & Electric Company (U 902 G) for Renewable Natural Gas Tariff, February 29, 2019, https://tinyurl.com/ex6zsmyv.
6. Sammy Roth, "The Fossil Fuel Industry Wants You to Believe It's Good for People of Color," *Los Angeles Times,* November 23, 2020.

2. THE ARSENAL

1. Tobacco Industry Meeting, New York, December 14, 1953, Philip Morris Records, Master Settlement Agreement, https://www.industrydocuments.ucsf.edu/docs/fgkj0191.
2. Peter Galison and Robert Proctor, "Agnotology in Action: A Dialogue," Chapter 2 in *Science and the Production of Ignorance: When the Quest for Knowledge Is Thwarted,* edited by Janet Kourany and Martin Carrier (Cambridge: MIT Press, 2020).
3. Carey Gillam, *Whitewash: The Story of a Weed Killer, Cancer, and the Corruption of Science* (Washington, DC: Island Press, 2017).
4. Max Boykoff and Justin Farrell, "Climate Change Countermovement Organizations and Media Attention in the United States," in *Climate Change Denial and Public Relations: Strategic Communication and Interest Groups in Climate Inaction,* edited by Núria Almiron and Jordi Xifra (Oxfordshire: Routledge, 2019).
5. Robert N. Proctor, *Golden Holocaust: Origins of the Cigarette Catastrophe and the Case for Abolition* (Berkeley: University of California Press, 2012).
6. Western Union telegram sent by Paul M. Hahn to Edward A. Darr, March 31, 1953, Philip Morris Records, Master Settlement Agreement, https://www.industrydocuments.ucsf.edu/docs/zrhw0181.
7. Thomas Whiteside, "A Cloud of Smoke," *New Yorker,* November 30, 1963.
8. Steve Koll, *Private Empire: ExxonMobil and American Power* (New York: Penguin, 2012).
9. Clifford Krauss, "Exxon Mobil's Chief Says It Is 'Supportive of Zero Emission Goals,'" *New York Times,* March 3, 2021.

10. Elliott Negin, "ExxonMobil Claims Shift on Climate but Continues to Fund Climate Science Deniers," *Union of Concerned Scientists/The Equation,* October 22, 2020.
11. Lucas Reilly, "The Most Important Scientist You've Never Heard Of," *Mental Floss,* May 17, 2017.
12. Meredith Hoffman, "Leading Scientists Tell the Nation's Museums to Sever Ties with the Koch Brothers," *Vice News,* March 24, 2015.
13. Koll, *Private Empire.*
14. Karen Miller Russell, *Promoting Monopoly: AT&T and the Politics of Public Relations, 1876–1941* (New York: Peter Lang, 2020).
15. Adam Curtis, "The Curse of Tina." *Adam Curtis: The Medium and the Message,"* (BBC blog), September 13, 2011.
16. Marc Gunter, "Edelman Loses Executives and Clients over Climate Change Stance," *Guardian,* July 7, 2015.
17. Ameet Sachdev, "PR Executive Sets Off Firestorm with Proposal to Discredit Madison County Court System," *Chicago Tribune,* May 28, 2011.
18. Lucy Michaels and Katharine Ainger, "The Climate Smokescreen: The Public Relations Consultancies Working to Obstruct Greenhouse Gas Emissions Reductions in Europe—A Critical Approach," in *Climate Change Denial and Public Relations: Strategic Communication and Interest Groups in Climate Inaction,* edited by Núria Almiron and Jordi Xifra (Oxfordshire: Routledge, 2019).
19. Karen Miller, *The Voice of Business: Hill & Knowlton and Postwar Public Relations* (Chapel Hill: University of North Carolina Press, 1999), 129.
20. Miller, *The Voice of Business.*
21. Miller, *The Voice of Business.*
22. Tobacco Industry Meeting, New York, December 14, 1953, Philip Morris Records, Master Settlement Agreement.
23. John W. Hill et al., "Smoking Health and Statistics: The Story of the Tobacco Accounts," http://legacy.library.ucsf.edu/tid/ksn33c00.
24. Gerald Markowitz and David Rosner, *Deceit and Denial: The Deadly Politics of Industrial Pollution* (Berkeley: University of California Press, 2002).
25. Sachdev, "PR Executive Sets Off Firestorm with Proposal to Discredit Madison County Court System."
26. Hill et al., "Smoking Health and Statistics: The Story of the Tobacco Accounts."

27. Craig Silverman, Jane Lytvynenko, and William Kung, "Disinformation for Hire: How a New Breed of PR firms Is Selling Lies Online," *BuzzFeed News,* January 6, 2020.
28. Sharon Beder, "Public Relations' Role in Manufacturing Artificial Grass Roots Coalitions," *Public Relations Quarterly* 43 (1998): 20–23.
29. Beder, "Public Relations' Role in Manufacturing Artificial Grass Roots Coalitions."
30. Clare Howard, "Pest Control: Syngenta's Secret Campaign to Discredit Atrazine's Critics," *100 Reporters,* June 17, 2013.
31. Jonathan Hiskes, "Lobby Firm Forges Anti-Climate-Bill Letters from Hispanic Group and NAACP," *Grist,* August 1, 2009.
32. Jim Snyder, "Coal Group, Grassroots Firm Knew of Forged Letters Before Climate Vote," *Hill,* October 29, 2009.
33. Proctor, *Golden Holocaust.*
34. Stanton A. Glantz et al., eds., *The Cigarette Papers* (Berkeley: University of California Press, 1996).
35. Letter from E. Weidlein, Director, Mellon Institute of Industrial Research, University of Pittsburgh, to R. Hitchins, President, American Refractories Institute (January 21, 1935), enclosed in letter from W. G. Hazard, Owens-Illinois Glass Co., to L. R. Thompson, U.S. Public Health Service (March 21, 1935) (on file with Mellon Institute National Archives), cited in Jenny White and Lisa A. Bero, "Corporate Manipulation of Research: Strategies Are Similar Across Five Industries, *Stanford Law & Policy Review* 21 (2010): 105–34.
36. "Donors Trust Principled Giving," https://www.donorstrust.org/who-we-are/mission-principles.
37. Robert Brulle, "Institutionalizing Delay: Foundation Funding and the Creation of U.S. Climate Change Counter-Movement Organizations," *Climatic Change* 122 (2014): 681–94.
38. Howard, "Pest Control: Syngenta's Secret Campaign to Discedit Atrazine's Critics."
39. Jim Dwyer, "What Happened to Jane Mayer When She Wrote About the Koch Brothers," *New York Times,* January 26, 2016.
40. Lauren Kelley, "Author Jane Mayer on How the Koch Brothers Have Changed America," *Rolling Stone,* February 14, 2016.
41. Eric Lipton, "Food Industry Enlisted Academics in G.M.O. Lobbying War, Emails Show," *New York Times,* September 5, 2015.
42. Andreas Lundh et al., "Industry Sponsorship and Research Outcome," *Cochrane Database Systematic Reviews* 2 (2017): MR000033.

43. David S. Ludwig, Lawrence H. Kushi, and Steven B. Heymsfield, "Conflicts of Interest in Nutrition Research," *Journal of the American Medical Association* 320 (2018): 93.
44. Maira Bes-Rastrollo et al., "Financial Conflicts of Interest and Reporting Bias Regarding the Association Between Sugar-Sweetened Beverages and Weight Gain: A Systematic Review of Systematic Reviews," *PLOS Medicine* 10 (2013): e1001578.
45. Lola Adekunle et al., "Association Between Financial Links to Indoor Tanning Industry and Conclusions of Published Studies on Indoor Tanning: Systematic Review," *British Medical Journal* 368 (2020): m7.
46. Naomi Oreskes and Erik M. Conway, *Merchants of Doubt: How a Handful of Scientists Obscured the Truth on Issues from Tobacco Smoke to Global Warming* (London: Bloomsbury, 2010).
47. Curtis, "The Curse of Tina."
48. Curtis, "The Curse of Tina."
49. RAND's Heath Insurance Experiment (HIE), https://www.rand.org/.
50. Jane Mayer, *Dark Money: The Hidden History of the Billionaires Behind the Rise of the Radical Right* (New York: Doubleday, 2016).
51. Riley E. Dunlap and Aaron McCright, "Challenging Climate Change: The Denial Countermovement," in *Climate Change and Society: Sociological Perspectives,* edited by Riley E. Dunlap and Robert J. Brulle (New York: Oxford University Press, 2015), 300–332.
52. Testimony of Mario Lewis, Jr., Vice President for Policy and Coalitions, Competitive Enterprise Institute, before the Committee on Small Business June 4, 1998, Hearing on the Kyoto Protocol.
53. Peter J. Jacques, Riley E. Dunlap, and Mark Freeman, "The Organisation of Denial: Conservative Think Tanks and Environmental Scepticism," *Environmental Politics* 17:3 (2008): 349–85.
54. Justin Gillis, "Clouds' Effect on Climate Change Is Last Bastion for Dissenters," *New York Times,* April 30, 2012.
55. Neela Banerjee, "How Big Oil Lost Control of Its Climate Misinformation Machine," *Inside Climate News,* December 22, 2017.
56. David Michaels, *The Triumph of Doubt: Dark Money and the Science of Deception* (Oxford: Oxford University Press, 2019).
57. David Michaels and Celeste Monforton, "Manufacturing Uncertainty: Contested Science and the Protection of the Public's Health and Environment," *American Journal of Public Health* 95, Suppl. 1 (2005): S39–48.

58. Myron Levin and Paul Feldman, "Big Companies in Legal Scrapes Turn to Science-for-Hire Giant Exponent," *Business Ethics,* December 13, 2016.
59. Exponent, "About Us," https://www.exponent.com/about/about-us.
60. Paul D. Boehm and Peter D. Carragher, "Location of Natural Oil Seep and Chemical Fingerprinting Suggest Alternative Explanation for Deep Sea Coral Observations," *Proceedings of the National Academy of Sciences* 109 (2012): e2647.
61. Levin and Feldman, "Big Companies in Legal Scrapes Turn to Science-for-Hire Giant Exponent."
62. Mary M. Murphy et al., "Fresh and Lean Pork Are Substantial Sources of Key Nutrients When These Products Are Consumed by Adults in the United States," *Nutrition Research* 31 (2011): 776–83.

CASE: EXPERT OPPOSITION TO DIVESTMENT

1. Matthew Taylor, "Half of UK Universities Have Committed to Divest from Fossil Fuel," *Guardian,* January 13, 2020.
2. Ivo Welch, "Why Divestment Fails," *New York Times,* May 9, 2014.
3. Paul Tice, "Fossil-Fuel Divestment Is Futile," *Wall Street Journal,* May 29, 2018.
4. Paul Tice, "On Climate, the Kids Are All Wrong," *Wall Street Journal,* March 12, 2019.
5. IPAA, "Cornell fact sheet," Divestment Facts, 2015, http://divestment facts.com.

3. RECRUITING UNIVERSITY EXPERTS

1. Pew Research Center, "Science and Scientists Held in High Esteem Across Global Publics," September 2020.
2. Samuel V. Bruton and Donald F. Sacco, "What's It to Me? Self-interest and Evaluations of Financial Conflicts of Interest," *Research Ethics* 14 (2017): 1–17.
3. Robert N. Proctor, *Golden Holocaust: Origins of the Cigarette Catastrophe and the Case for Abolition* (Berkeley: University of California Press, 2012).
4. Anna Maria Barry-Jester, "Amid Teen Vaping 'Epidemic,' Juul Taps Addiction Expert as Medical Director," *Philly Voice,* July 22, 2019.

5. Annie Waldman and David Armstrong, "Many Public Universities Refuse to Reveal Professors' Conflicts of Interest," *Chronicle of Higher Education,* December 6, 2019.
6. Steven Pearlstein, "I Teach at George Mason. The Kochs Didn't Cause Our Ideology Problem," *Washington Post,* May 16, 2018.
7. Allan M. Brandt, "Inventing Conflicts of Interest: A History of Tobacco Industry Tactics," *American Journal of Public Health* 102 (2012): 63–71.
8. Naomi Oreskes and Erik M. Conway, *Merchants of Doubt: How a Handful of Scientists Obscured the Truth on Issues from Tobacco Smoke to Global Warming* (London: Bloomsbury, 2010).
9. Brandt, "Inventing Conflicts of Interest: A History of Tobacco Industry Tactics."
10. Proctor, *Golden Holocaust.*
11. Riley E. Dunlap and Aaron McCright, "Challenging Climate Change: The Denial Countermovement," in *Climate Change and Society: Sociological Perspectives,* edited by Riley E. Dunlap and Robert J. Brulle (New York: Oxford University Press, 2015), 300–332.
12. John H. Cushman, Jr., "Industrial Group Plans to Battle Climate Treaty," *New York Times,* April 26, 1998.
13. Laura Krantz, "Harvard Professor Failed to Disclose Connection," *Boston Globe,* October 1, 2015.
14. Sam Bloch, "Sorry, Alexandria Ocasio-Cortez but 'Farting Cows' Aren't the Problem," *The Counter,* March 7, 2019.
15. Carey Gillam, *Whitewash: The Story of a Weed Killer, Cancer, and the Corruption of Science* (Washington, DC: Island Press, 2017).
16. John Hocevar, "Ray Hilborn: Overfishing Denier," Greenpeace.org, May 12, 2016.
17. https://rayhblog.wordpress.com/.
18. Proctor, *Golden Holocaust.*
19. DairyBusiness News Team DP, "Air Quality Expert Dr. Frank Mitloehner to Debunk Environmental Myths at 2018 Stakeholders Summit," *Dairy Business,* March 14, 2018.
20. Sylvia Wright, "Don't Blame Cows for Climate Change," UC Davis press release, December 7, 2009, https://www.ucdavis.edu/news/don%E2%80%99t-blame-cows-climate-change.
21. Robert N. Proctor, "'Everyone Knew but No One Had Proof': Tobacco Industry Use of Medical History Expertise in US Courts, 1990–2002," *Tobacco Control* 15, Suppl. 4 (2006): iv117—iv125.

22. Roland C. Clement, "The Pesticides Controversy," *Boston College Environmental Affairs Law Review* 2 (1972): 445–68.
23. Paul Brodeur, "The Asbestos Industry on Trial," *New Yorker,* June 17, 1985, 45.
24. Danny Hakim, "Scientists Loved and Loathed by an Agrochemical Giant," *New York Times,* December 31, 2016.
25. Lisa Song, "Scientific Journals Alerted to Fossil Fuel Funding of Contrarian Climate Studies," *Inside Climate News,* February 23, 2015.
26. David Michaels, *The Triumph of Doubt: Dark Money and the Science of Deception* (Oxford: Oxford University Press, 2019).
27. Elham Shabahat, "'Antithetical to Science': When Deep-Sea Research Meets Mining Interests," *Mongabay,* October 4, 2021.
28. Frank Davidoff et al., "Sponsorship, Authorship, and Accountability," *The Journal of the American Medical Association* 286 (2001): 1232.
29. Hakim, "Scientists Loved and Loathed by an Agrochemical Giant."
30. Hal Bernton, "Greenpeace Files Complaint About UW Fishery Professor," *Seattle Times,* May 11, 2016.
31. Anahad O'Connor, "Coca-Cola Funds Scientists Who Shift Blame for Obesity Away from Bad Diets," *New York Times,* August 9, 2015.
32. Richard E. L. Rogers and James R. Kemp, "Imidacloprid, Potatoes, and Honey Bees in Atlantic Canada: Is There a Connection?," *Bulletin of Insectology* 56 (2003): 83–88.
33. Sally Casswell et al., "How the Alcohol Industry Relies on Harmful Use of Alcohol and Works to Protect Its Profits," *Drug and Alcohol Review* 35 (2016): 661–64.
34. Sean Cowlishaw and Samantha L. Thomas, "Industry Interests in Gambling Research: Lessons Learned from Other Forms of Hazardous Consumption," *Addictive Behaviors* 78 (2018): 101–6.
35. Bruton and Sacco, "What's It to Me? Self-interest and Evaluations of Financial Conflicts of Interest."
36. Waldman and Armstrong, "Many Public Universities Refuse to Reveal Professors' Conflicts of Interest."
37. Stephanie Saul, "Conflict on the Menu," *New York Times,* February 16, 2008.
38. Cass R. Sunstein et al., *Punitive Damages: How Juries Decide* (Chicago: University of Chicago Press, 2002).
39. David Stuckler, Gary Ruskin, and Martin McKee, "Complexity and Conflicts of Interest Statements: A Case-Study of Emails Exchanged Between Coca-Cola and the Principal Investigators of the Interna-

tional Study of Childhood Obesity, Lifestyle and the Environment (ISCOLE)," *Journal of Public Health Policy* 39 (2018): 49–56.

40. Lisa Bero, "Ten Tips for Spotting Industry Involvement in Science Policy," *Tobacco Control* 28 (2019): 1–2.
41. Lisa Bero and Quinn Grundy, "Why Having a (Nonfinancial) Interest Is Not a Conflict of Interest," *PLOS Biology* 14 (2016): e2001221.
42. Proctor, *Golden Holocaust*.
43. Paul Thacker, "How an Ethically Challenged Researcher Found a Home at the University of Miami," *Forbes*, September 13, 2011.
44. Colin Buxton et al., "Super Trawler Gone, but Is Fisheries Policy in Trouble?," *Conversation*, March 12, 2013.
45. Nicholas Kohler, "Where's the Beef? Scientist Takes a Second Look at UN Numbers That Have Led Many Environmentalists to Forego Meat," *Macleans*, March 30, 2010.
46. Dennis F. Thomson, "Understanding Financial Conflicts of Interest," *New England Journal of Medicine* 329 (1993): 573–76.
47. Bruton and Sacco, "What's It to Me? Self-interest and Evaluations of Financial Conflicts of Interest."
48. Thacker, "How an Ethically Challenged Researcher Found a Home at the University of Miami."
49. "Managing Conflicts of Interest in the NHS," https://www.england.nhs.uk/ourwork/coi/.
50. Thacker, "How an Ethically Challenged Researcher Found a Home at the University of Miami."
51. Waldman and Armstrong, "Many Public Universities Refuse to Reveal Professors' Conflicts of Interest."
52. "*Nature* Competing Interests," https://www.nature.com/.
53. Proctor, *Golden Holocaust*.
54. *PLOS One* Competing Interests, https://journals.plos.org/plosone/s/-interests.
55. Margaret Sullivan, "Hidden Interests, Closer to Home," *New York Times*, September 20, 2014.
56. Andreas Lundh et al., "Industry Sponsorship and Research Outcome," *Cochrane Database Systematic Reviews* 2 (2017): MR000033.
57. Jennifer Jacquet, "Guilt and Shame in U.S. Climate Change Communication," *The Oxford Encyclopedia of Climate Change Communication*, 2017.
58. Melody Petersen, "Undisclosed Financial Ties Prompt Reprove of Doctor," *New York Times*, August 3, 2003.

59. Jeff Tollefson, "Earth Science Wrestles with Conflict-of-Interest Policies," *Nature* 522 (2015): 403–4.
60. Bernton, "Greenpeace Files Complaint About UW Fishery Professor."
61. Donald I. Siegel et al., "Methane Concentrations in Water Wells Unrelated to Proximity to Existing Oil and Gas Wells in Northeastern Pennsylvania," *Environmental Science and Technology* 49 (2015): 4106–12.
62. Neela Banerjee, "Fracking Study on Water Contamination Under Ethics Review," *Inside Climate News,* April 6, 2015.
63. Siegel et al., "Methane Concentrations in Water Wells Unrelated to Proximity to Existing Oil and Gas Wells in Northeastern Pennsylvania."
64. Jack Kaskey, "How Monsanto Mobilized Academics to Pen Articles Supporting GMOs," *Bloomberg,* October 2, 2015.
65. Eric Lipton, "Food Industry Enlisted Academics in G.M.O. Lobbying War, Emails Show," *New York Times,* September 5, 2015.
66. Saul, "Conflict on the Menu."
67. Frank Mitloehner, comment on Jillian Fry, "Unsupported Claims About Livestock and Climate Change in the Media," *Center for a Livable Future* (blog), March 29, 2010.
68. Bernton, "Greenpeace Files Complaint About UW Fishery Professor."
69. Bernton, "Greenpeace Files Complaint About UW Fishery Professor."
70. Roland C. Clement, "The Pesticides Controversy," *Boston College Environmental Affairs Law Review* 2 (1972): 445–68.
71. Colleen Flaherty, "Court Sides with 'The New York Times' in Professor's Defamation Case," *Inside Higher Ed,* March 1, 2019.
72. Charles Ornstein and Katie Thomas, "Top Cancer Researcher Fails to Disclose Corporate Financial Ties in Major Research Journals," *New York Times,* September 8, 2018.
73. Proctor, *Golden Holocaust.*
74. Geoff Dyer, *Selfish, Shallow, and Self-Absorbed: Sixteen Writers on the Decision Not to Have Kids* (London: Picador, 2015).
75. Brodeur, "The Asbestos Industry on Trial," 82.
76. Tyrone Hayes, personal communication.
77. Dashka Slater, "The Frog of War," *Mother Jones,* January–February 2012.
78. Clare Howard, "Pest Control: Syngenta's Secret Campaign to Discredit Atrazine's Critics," *100 Reporters,* June 17, 2013. Also see: https://tinyurl.com/4896e3bf.

79. Rachel Aviv, "A Valuable Reputation," *New Yorker,* February 10, 2014.
80. Slater, "The Frog of War."
81. Howard, "Pest Control: Syngenta's Secret Campaign to Discredit Atrazine's Critics."

CASE: HOW THE FOOD INDUSTRY COMMUNICATES ON OBESITY

1. The Center for Consumer Freedom, "About Us," https://www.consumerfreedom.com/about/.
2. Laura Nixon et al., "'We're Part of the Solution': Evolution of the Food and Beverage Industry's Framing of Obesity Concerns Between 2000 and 2012," *American Journal of Public Health* 105 (2015): 2228–36.
3. Jaime Holguin, "Battle of the Widening Bulge," *CBS News,* August 8, 2002.

4. STRATEGIC COMMUNICATION

1. Lee Fang, "The Playbook for Poisoning the Earth," *Intercept,* January 18, 2020.
2. John W. Hill et al., "Smoking Health and Statistics: The Story of the Tobacco Accounts," http://legacy.library.ucsf.edu/tid/ksn33c00.
3. Mi-Kyung Hong and Lisa A. Bero, "How the Tobacco Industry Responded to an Influential Study of the Health Effects of Secondhand Smoke," *The British Medical Journal* 325 (2002): 1413.
4. Charles Duhigg, "Debating How Much Weed Killer Is Safe in Your Water Glass," *New York Times,* August 22, 2009.
5. Paul Towers, "Syngenta Hired Guns Attack New Documentary," *PR Watch,* May 11, 2012.
6. Rachel Aviv, "A Valuable Reputation," *New Yorker,* February 10, 2014.
7. Email from Jessica Adelman to Steven Goldsmith, August 28, 2009, https://www.sourcewatch.org/images/e/e9/Syn_email_ACSH_Is_Great_Weapon.pdf.
8. Robert J. Brulle, "Denialism: Organized Opposition to Climate Change Action in the United States," in *Handbook of U.S. Environmental Policy,* edited by David M. Konisky (Cheltenham: Edward Elgar, 2020).
9. Eliza Barclay, "Monsanto Hired This Guy to Help It Win Over Millennials," *The Salt,* National Public Radio, October 29, 2014.

10. "On the Farm, STEM," https://www.onthefarmstem.com/participants/2019.
11. Susan J. Ainsworth, "Dow Chemical Partners with Chemistry Teacher Group," *Chemical & Engineering News,* January 20, 2015.
12. "Science Coaches Spotlight," *American Association of Chemistry Teachers,* December 11, 2019.
13. Neela Banerjee, "Fracking Study on Water Contamination Under Ethics Review," *Inside Climate News,* April 6, 2015.
14. Donald Siegel, "'Shooting the Messenger': Some Reflections on What Happens Doing Science in the Public Arena," *Hydrological Processes* 30 (2016): 830–32.
15. Desmond Butler and Juliet Eilperin, "The Anti-Greta: A Conservative Think Tank Takes on the Global Phenomenon," *Washington Post,* February 23, 2020.
16. Justin Scheck, Eliot Brown, and Ben Foldy, "Environmental Investing Frenzy Stretches Meaning of 'Green,'" *Wall Street Journal,* June 24, 2021.
17. Thomas Whiteside, "A Cloud of Smoke," *New Yorker,* November 30, 1963.
18. "Centuries-Old Smoking/Health Controversy Continues," January 1968, R.J. Reynolds Records, Master Settlement Agreement, https://www.industrydocuments.ucsf.edu/docs/hgxf0083.
19. Steve Koll, *Private Empire: ExxonMobil and American Power* (New York: Penguin, 2012), 338.
20. Willy Blackmore, "Monsanto, Blamed for Killing Monarchs, Donates Millions to Save the Butterflies," *Take Part,* April 1, 2015.
21. Sheera Frenkel et al., "Delay, Deny and Deflect: How Facebook's Leaders Fought Through Crisis," *New York Times,* November 16, 2018.
22. Whiteside, "A Cloud of Smoke."
23. Robert N. Proctor, *Golden Holocaust: Origins of the Cigarette Catastrophe and the Case for Abolition* (Berkeley: University of California Press, 2012).
24. Coral Davenport and Mark Landler, "Trump Administration Hardens Its Attack on Climate Science," *New York Times,* May 27, 2019.
25. Arthur Neslen, "Coal Lobby Says Industry 'Will Be Hated like Slave Traders' After COP21," *Guardian,* December 15, 2015.
26. Clare Howard, "Pest Control: Syngenta's Secret Campaign to Discredit Atrazine's Critics," *100 Reporters,* June 17, 2013.

27. Terry Mcalister, "BP Rebrands on a Global Scale," *Guardian,* July 24, 2000.
28. "CropLife America," SourceWatch, https://www.sourcewatch.org/index.php/CropLife_America.
29. "Indoor Tanning Association Settles FTC Charges That It Deceived Consumers About Skin Cancer Risks from Tanning," Federal Trade Commission, January 26, 2010.
30. David Michaels and Celeste Monforton, "Manufacturing Uncertainty: Contested Science and the Protection of the Public's Health and Environment," *American Journal of Public Health* 95, Suppl. 1 (2005): S39–S48.
31. American Suntanning Association, "Indoor Tanning Salon Leaders Announce Formation of American Suntanning Association," December 18, 2012.
32. American Suntanning Association, "Indoor Tanning Salon Leaders Announce Formation of American Suntanning Association."
33. Charles Ornstein and Tracy Weber, "American Pain Foundation Shuts Down as Senators Launch Investigation of Prescription Narcotics," *ProPublica,* May 8, 2012.
34. Paul Brodeur, "The Asbestos Industry on Trial," *New Yorker,* June 17, 1985, 76.
35. David J. Lewinter, "RJR Nabisco 1996 Annual Meeting," April 18, 1996, Philip Morris Records, Master Settlement Agreement, https://www.industrydocuments.ucsf.edu/docs/nsyh0018.
36. Nixon et al., "'We're Part of the Solution': Evolution of the Food and Beverage Industry's Framing of Obesity Concerns Between 2000 and 2012."
37. Proctor, *Golden Holocaust.*
38. Alexander Michael Petersen, Emmanuel M. Vincent, and Anthony LeRoy Westerling, "Discrepancy in Scientific Authority and Media Visibility of Climate Change Scientists and Contrarians," *Nature Communications* 10 (2019): 3502.
39. https://vaporvoice.net/about-us/.
40. National Cattlemen's Beef Association, "Study Confirms US Beef Industry Is the Most Sustainable in the World," press release, April 5, 2021.
41. Carey Gillam, *Whitewash: The Story of a Weed Killer, Cancer, and the Corruption of Science* (Washington, DC: Island Press, 2017).
42. Aviv, "A Valuable Reputation."

43. Beef. It's What's for Dinner, "Why Taxing Beef Isn't the Answer," *Medium,* March 27, 2018.
44. Michaels and Monforton, "Manufacturing Uncertainty: Contested Science and the Protection of the Public's Health and Environment."
45. *American Journal of Potato Research,* https://www.springer.com/journal/12230.
46. Jie Jenny Zou, "Brokers of Junk Science?," The Center for Public Integrity, February 18, 2016.
47. Gillam, *Whitewash.*
48. Sharon Lerner, "The Teflon Toxin," *Intercept,* August 11, 2015.
49. Rachel Wetts, "In Climate News, Statements from Large Businesses and Opponents of Climate Action Receive Heightened Visibility," *Proceedings of the National Academy of Sciences* 117 (2020): 19054–60.
50. White House Writers Group, "Syngenta Crop Protection PR/Advertising Agency RFP," September 2009, https://tinyurl.com/2pe39usj.
51. Chris Lischewski, Shue Wing Chan, and In-Soo Cho, "Greenpeace vs. the Tuna Sandwich," *Wall Street Journal,* November 8, 2011.
52. Anne Landman and Stanton A. Glantz, "Tobacco Industry Efforts to Undermine Policy-Relevant Research," *American Journal of Public Health* 99 (2009): 45–48.
53. Proctor, *Golden Holocaust.*
54. Rachel Koning Beals, "Oil Giant Shell Tries to Hand Off the Climate-Change Fight to Consumers and Gets Roasted by AOC, Greta Thunberg and Thousands More," *MarketWatch,* November 3, 2020.
55. "Rachel Carson," *American Experience,* PBS, 2017.
56. Proctor, *Golden Holocaust.*
57. Sander van Der Linden et al., "Inoculating the Public Against Misinformation About Climate Change," *Global Challenges* 1 (2017): 1600008.
58. Chris Mooney, *The Republican War on Science* (New York: Basic Books, 2005).
59. Mary Mangan, "Hogwash! A Review of *Whitewash* by Carey Gillam," *Biofortified* (blog), February 14, 2018.
60. https://s3.documentcloud.org/documents/686406/100reporters-syngenta-clare-howard-investigation.pdf.
61. Dashka Slater, "The Frog of War," *Mother Jones,* January–February 2012.
62. Karen Miller, *The Voice of Business: Hill & Knowlton and Postwar Public Relations* (Chapel Hill: University of North Carolina Press, 1999).

63. Proctor, *Golden Holocaust.*
64. Gillam, *Whitewash.*
65. Tzeporah Berman, "Canada's Most Shameful Environmental Secret Must Not Remain Hidden," *Guardian,* November 14, 2017.

5. CHALLENGE THE PROBLEM

1. Benjamin Franta, "Early Industry Knowledge of CO2 and Global Warming," *Nature Climate Change* 8 (2018): 1024–25.
2. Neela Banerjee, Lisa Song, and David Hasemyer, "Exxon's Own Research Confirmed Fossil Fuels' Role in Global Warming Decades Ago," *Inside Climate News,* September 16, 2015.
3. ExxonMobil, "Understanding the #ExxonKnew Controversy," February 10, 2021, https://corporate.exxonmobil.com.
4. Paul Brodeur, "The Asbestos Industry on Trial," *New Yorker,* June 17, 1985, 76.
5. Matthew L. Wald, "Pro-Coal Ad Campaign Disputes Warming Idea," *New York Times,* July 8, 1991.
6. Naomi Oreskes, Erik M. Conway, and Matthew Shindell, "From Chicken Little to Dr. Pangloss: William Nierenberg, Global Warming, and the Social Deconstruction of Scientific Knowledge," *Historical Studies in the Natural Sciences* 38 (2008): 109–52.
7. Lee Raymond, "Speech at the World Petroleum Congress in Beijing, China," October 13, 1997, http://www.climatefiles.com/.
8. Otto Pohl, "Challenge to Fishing: Keep the Wrong Species Out of Its Huge Nets," *New York Times,* July 29, 2003.
9. Bennet I. Omalu et al., "Chronic Traumatic Encephalopathy in a National Football League Player," *Neurosurgery* 57 (2005): 128–34.
10. Ira R. Casson, Elliot J. Pellman, and David C. Viano, "Chronic Traumatic Encephalopathy in a National Football League Player," *Neurosurgery* 58 (2006): E1003.
11. Annie Kleykamp and Helen Redmond, "COVID and Vaping: A Perfect Storm of Misleading Science and Media," *Filter,* October 14, 2020.
12. Letter from Leonard A. Miller, Attorney, Swidler & Berlin, to Thomas Borelli, Director of Science and Environmental Policy, Philip Morris, February 2, 1993, https://www.industrydocuments.ucsf.edu/tobacco/docs/#id=hygw0019.
13. Mayanna Lahsen, "Anatomy of Dissent: A Cultural Analysis of Climate Skepticism," *American Behavioral Scientist* 57 (2013): 732–53.

14. Steven E. Koonin, "Climate Science Is Not Settled," *Wall Street Journal,* September 19, 2014.
15. David Michaels and Celeste Monforton, "Manufacturing Uncertainty: Contested Science and the Protection of the Public's Health and Environment," *American Journal of Public Health* 95, Suppl. 1 (2005): S39–S48.
16. George R. Tilton, *Clair Cameron Patterson 1922–1995: A Biographical Memoir* (Washington, DC: National Academies Press, 1998), http://www.nasonline.org/publications/biographical-memoirs/memoir-pdfs/patterson-clair-c.pdf.
17. Alex Berezow, "Poverty, Not Climate Change, Remains World's Deadliest Problem," American Council on Science and Health, February 27, 2018.
18. Ray Hilborn, "Policy: Marine Biodiversity Needs More Than Protection," *Nature* 535 (2016): 224–26.
19. Vasile Stanescu, "Cowgate: Meat Eating and Climate Change Denial," Chapter 10 in *Climate Change Denial and Public Relations: Strategic Communication and Interest Groups in Climate Inaction,* edited by Núria Almiron and Jordi Xifra (Oxfordshire: Routledge, 2019).
20. Justin Scheck, Eliot Brown, and Ben Foldy, "Environmental Investing Frenzy Stretches Meaning of 'Green,'" *Wall Street Journal,* June 24, 2021.
21. This is a quote from a slide presented by Dr. Ray Hilborn in his testimony before Congress, Committee on Natural Resources, on September 11, 2013, about the reauthorization of the Magnuson-Stevens Fishery Conservation and Management Act.
22. Robert Kearney, "Australia's Out-dated Concern over Fishing Threatens Wise Marine Conservation and Ecologically Sustainable Seafood Supply," *Open Journal of Marine Science* 3 (2013): 55–61.
23. Robert N. Proctor, *Golden Holocaust: Origins of the Cigarette Catastrophe and the Case for Abolition* (Berkeley: University of California Press, 2012).
24. Oliver Burkeman, "Memo Exposes Bush's New Green Strategy," *Guardian,* March 3, 2003.
25. *Toxic Sludge Is Good For You: The Public Relations Industry Unspun,* 2003, http://bufvc.ac.uk/dvdfind/index.php/title/av35076.
26. Barbara Freese, *Industrial-Strength Denial: Eight Stories of Corporations Defending the Indefensible, from the Slave Trade to Climate Change* (Berkeley: University of California Press, 2020).

27. Rainer Froese and Alexander Proelss, "Evaluation and Legal Assessment of Certified Seafood," *Marine Policy* 36 (2012): 1284–89.
28. Daniel Cressey, "Seafood Labelling Under Fire," *Nature,* May 11, 2012.
29. David J. Agnew et al., "Rebuttal to Froese and Proelss 'Evaluation and legal assessment of certified seafood,'" *Marine Policy* 38 (2013): 551–53.
30. Steven Koonin, guest lecture in Roy Radner's Business and the Environment class at NYU, April 2, 2015.
31. Whole Foods Market, "Responsibly Farmed Salmon," YouTube, March 10, 2014.
32. Boris Worm et al., "Rebuilding Global Fisheries," *Science* 325 (2009): 578–85.
33. Worm et al., "Rebuilding Global Fisheries."
34. Valerio Gennaro and Lorenzo Tomatis, "Business Bias: How Epidemiologic Studies May Underestimate or Fail to Detect Increased Risks of Cancer and Other Diseases," *International Journal of Occupational and Environmental Health* 11 (2005): 356–59.
35. Alan Schwarz, Walt Bogdanich, and Jacqueline Williams, "N.F.L.'s Flawed Concussion Research and Ties to Tobacco Industry," *New York Times,* March 24, 2016.
36. Canadian Forest Industries, "Counting Caribou in Ontario," *Wood Business,* December 18, 2015.
37. Julee J. Boan et al., "From Climate to Caribou: How Manufactured Uncertainty Is Affecting Wildlife Management," *Wildlife Society Bulletin* 42 (2018): 366–81.
38. Ontario Forest Industries Association, "Counting Caribou: How Did Canada's Most Populous Ungulate End Up on Ontario's Endangered Species List?," 2015.
39. https://www.documentcloud.org/documents/6509361-DVE-Bayer-BeeInformed.html, cited in Lee Fang, "The Playbook for Poisoning the Earth," *Intercept,* January 18, 2020.
40. Daniel Pauly, "Aquacalypse Now, *New Republic,* September 27, 2009.
41. David A. Kroodsma et al., "Tracking the Global Footprint of Fisheries," *Science* 359 (2018): 904–8.
42. Ed Yong, "Wait, So How Much of the Ocean Is Actually Fished?," *The Atlantic,* September 10, 2018.
43. Ricardo O. Amoroso et al., Comment on "Tracking the Global Footprint of Fisheries," *Science* 361 (2018): eaat6713.

44. Paul Brodeur, "The Asbestos Industry on Trial," *New Yorker,* June 17, 1985, 75.
45. Brodeur, "The Asbestos Industry on Trial."

CASE: QUESTIONING THE RELATIONSHIP BETWEEN VAPING AND COVID-19

1. Wei-jie Guan et al., "Clinical Characteristics of Coronavirus Disease 2019 in China," *New England Journal of Medicine* 382 (2020): 1708–20.
2. Jan Hoffman, "Smokers and Vapers May Be at Greater Risk for Covid-19," *New York Times,* April 9, 2020.
3. Shivani Mathar Gaiha, Jing Cheng, and Bonnie Halpern-Felsher, "Association Between Youth Smoking, Electronic Cigarette Use, and COVID-19," *Journal of Adolescent Health* 67 (2020): 519–23.
4. Katherine J. Wu, "Vaping Links to Covid Risks Are Becoming Clear," *New York Times,* September 4, 2020.
5. Joe G. Gitchell et al., "Bad Data and Bad Conclusions Will Lead to Bad Policy—Implausible Claims That Vaping Increases COVID-19 Risk for Youth and Young Adults," Qeios preprint.
6. Annie Kleykamp and Helen Redmond, "COVID and Vaping: A Perfect Storm of Misleading Science and Media, *Filter,* October 14, 2020.
7. https://filtermag.org/author/annie-kleykamp/.
8. Brad Rodu, "California Researchers Make Dramatic Claims About E-Cigarettes & Covid-19, but Fail to Disclose Minuscule Case Numbers," *Tobacco Truth* (blog), August 19, 2020.
9. John W. Hill, letter to Dr. Clarence Cook Little, Re: TIRC / SAB, July 15, 1954, https://www.industrydocuments.ucsf.edu/docs/fsfy0042.

6. CHALLENGE CAUSATION

1. Oliver Lazarus, Sonali McDermid, and Jennifer Jacquet, "The Climate Responsibilities of Industrial Meat and Dairy Producers," *Climatic Change* 165 (2021): 30.
2. David Michaels and Celeste Monforton, "Manufacturing Uncertainty: Contested Science and the Protection of the Public's Health and Environment," *American Journal of Public Health* 95, Suppl. 1 (2005): S39–S48.

3. Charles Duhigg, "Debating How Much Weed Killer Is Safe in Your Water Glass," *New York Times,* August 22, 2009.
4. Barbara Freese, *Industrial-Strength Denial: Eight Stories of Corporations Defending the Indefensible, from the Slave Trade to Climate Change* (Berkeley: University of California Press, 2020).
5. Sharon Lerner, "The Teflon Toxin," *Intercept,* August 11, 2015.
6. Thomas Whiteside, "A Cloud of Smoke," *New Yorker,* November 30, 1963, 105.
7. Eric Sachs and Bruce Chassy, email exchanges, September 13, 2011, obtained by U.S. Right to Know, cited in Carey Gillam, *Whitewash: The Story of a Weed Killer, Cancer, and the Corruption of Science* (Washington, DC: Island Press, 2017), 122.
8. Nina Teicholz, "EAT-Lancet Report Is One-Sided, Not Backed by Rigorous Science," *Nutrition Coalition,* January 29, 2019.
9. America's Heartland, "Native Bee Pollinators," YouTube, January 9, 2013.
10. Lee Fang, "The Playbook for Poisoning the Earth," *Intercept,* January 18, 2020.
11. Dale Jamieson, "Scientific Uncertainty and the Political Process," *Annals of the American Academy of Political and Social Science* 545 (1996): 35–43.
12. Whiteside, "A Cloud of Smoke."
13. Keum Ji Jung, Christina Jeon, and Sun Ha Jee, "The Effect of Smoking on Lung Cancer: Ethnic Differences and the Smoking Paradox," *Epidemiology and Health* 38 (2016): e2016060.
14. Henning Steinfeld et al., "Livestock's Long Shadow," Food and Agriculture Organization of the United Nations, 2006.
15. Sylvia Wright, "Don't Blame Cows for Climate Change," UC Davis press release, December 7, 2009.
16. Maurice E. Pitesky, Kimberly R. Stackhouse, and Frank M. Mitloehner, "Clearing the Air: Livestock's Contribution to Climate Change," *Advances in Agronomy* 103 (2009): 1–40.
17. Leo Hickman, "Do Critics of UN Meat Report Have a Beef with Transparency?" *The Guardian,* March 24, 2010.
18. Wright, "Don't Blame Cows for Climate Change."
19. Bill Sells, "What Asbestos Taught Me About Managing Risk," *Harvard Business Review,* March–April 1994.
20. Donald G. Cooley, "Who Says Smoking Causes Lung Cancer?," *True*

Magazine, July 1954, https://www.industrydocuments.ucsf.edu/docs/phdd0216.

21. Committee on the Analysis of Cancer Risks in Populations near Nuclear Facilities-Phase I; Nuclear and Radiation Studies Board; Division on Earth and Life Studies; National Research Council, *Analysis of Cancer Risks in Populations Near Nuclear Facilities: Phase I.* Washington (DC): National Academies Press (US); 2012 Mar 29. Available from: https://www.ncbi.nlm.nih.gov/books/NBK201996/doi:10.17226/13388.
22. Chris Mooney, *The Republican War on Science* (New York: Basic Books, 2005).
23. Josh Bloom, "BPA Alarmists Are Beating a Dead Monomer—Enough Already," *American Council on Science and Health,* December 30, 2019.
24. Peter Waldman, "Toxic Traces: New Questions About Old Chemicals," *Wall Street Journal,* July 25, 2005.
25. Pitesky, Stackhouse, and Mitloehner, "Clearing the Air: Livestock's Contribution to Climate Change."
26. Lazarus, McDermid, and Jacquet, "The Climate Responsibilities of Industrial Meat and Dairy Producers," 30.
27. Danny Hakim, "Scientists Loved and Loathed by an Agrochemical Giant," *New York Times,* December 31, 2016.
28. Mooney, *The Republican War on Science.*
29. Daniel Engber, "The Sugar Wars," *Atlantic,* January–February 2017.
30. Jerome F. Cole, "Old Paint, Not Gasoline, Is the Problem in Lead Poisoning," *New York Times,* August 23, 1982.
31. Lissy C. Friedman et al., "Tobacco Industry Use of Personal Responsibility Rhetoric in Public Relations and Litigation: Disguising Freedom to Blame as Freedom of Choice," *American Journal of Public Health* 105 (2015): 250–60.
32. Whiteside, "A Cloud of Smoke."
33. Mark Kaufman, "The Carbon Footprint Sham," *Mashable,* https://mashable.com/feature/carbon-footprint-pr-campaign-sham.
34. Barbara Freese, *Industrial-Strength Denial: Eight Stories of Corporations Defending the Indefensible, from the Slave Trade to Climate Change* (Berkeley: University of California Press, 2020).
35. Freese, *Industrial-Strength Denial.*
36. Lee Ferran and Russell Goldman, "SeaWorld Curator: Ponytail Likely Caused Fatal Killer Whale Attack," *ABC News,* February 24, 2010.

37. Albert Samaha and Katie J. M. Baker, "Smithfield Foods Is Blaming 'Living Circumstances in Certain Cultures' for One of America's Largest COVID-19 Clusters," BuzzFeed, April 20, 2020.
38. Sells, "What Asbestos Taught Me About Managing Risk."
39. Art Van Zee, "The Promotion and Marketing of OxyContin: Commercial Triumph, Public Health Tragedy," *American Journal of Public Health* 99 (2009): 221–27.
40. Alan Schwarz, "Dementia Risk Seen in Players in NFL Study," *New York Times,* September 30, 2009.
41. Judith L. Capper, "The Environmental Impact of Beef Production in the United States: 1977 Compared with 2007," *Journal of Animal Science* 89 (2011): 4249–61.
42. Judith L. Capper and Roger A Cady, "The Effects of Improved Performance in the U.S. Dairy Cattle Industry on Environmental Impacts Between 2007 and 2017," *Journal of Animal Science* 98 (2020): skz291.
43. Roland C. Clement, "The Pesticides Controversy," *Boston College Environmental Affairs Law Review* 2 (1972): 445–68.
44. Whiteside, "A Cloud of Smoke."
45. Jie Jenny Zou, "Brokers of Junk Science?," *Center for Public Integrity,* February 18, 2016.
46. Geoffrey Kabat, "Viewpoint: How Anti-GMO Activist-Journalist Carey Gillam Primes the Glyphosate Litigation Pump," *Genetic Literacy Project,* October 15, 2018.

CASE: BIAS IN THE EAT-*LANCET* COMMISSION

1. "The EAT-Lancet Commission on Food, Planet, Health," https://eatforum.org/eat-lancet-commission/.
2. David Garcia, Victor Galaz, and Stefan Daume, "EATLancet vs. Yes2meat: The Digital Backlash to the Planetary Health Diet," *Lancet* 394 (2019): 2153–54.
3. Nina Teicholz, "Majority of EAT-Lancet Authors (>80%) Favored Vegan/Vegetarian Diets," *Nutrition Coalition,* January 24, 2019.
4. Nina Teicholz, "EAT-Lancet Report Is One-Sided, Not Backed by Rigorous Science," *Nutrition Coalition,* January 29, 2019.
5. Diana Rodgers, "20 Ways EAT Lancet's Global Diet Is Wrongfully Vilifying Meat," *Sustainable Dish* (blog), January 17, 2019.

NOTES

7. CHALLENGE THE MESSENGER

1. Michael Warren, "Argentine Scientist Who Challenged Monsanto Dies," Associated Press, May 10, 2014.
2. Associated Press, "Andres Carrasco Dies; Argentine Neuroscientist Showed Monsanto's Glyphosate Damages Embryos," *Fox News*, May 10, 2014.
3. "Preliminary Recommendations for Cigarette Manufacturers," December 24, 1953, American Tobacco Records, Master Settlement Agreement, https://www.industrydocuments.ucsf.edu/docs/rsfy0137.
4. Karen Miller, *The Voice of Business: Hill & Knowlton and Postwar Public Relations* (Chapel Hill: University of North Carolina Press, 1999).
5. Forwarding Memorandum, May 19, 2008, Philip Morris Records, Master Settlement Agreement, https://www.industrydocuments.ucsf.edu/docs/ggkj0191.
6. Alex Berezow, "Glyphosate: NYT's Danny Hakim Is Lying to You," *American Council on Science and Health*, March 15, 2017.
7. Barbara Freese, *Industrial-Strength Denial: Eight Stories of Corporations Defending the Indefensible, from the Slave Trade to Climate Change* (Berkeley: University of California Press, 2020).
8. Geoff Dembicki, "How Exxon Silences Staff Alarmed by the Climate Crisis, According to a Former Employee," *Vice*, October 29, 2020.
9. Jane Mayer, *Dark Money: The Hidden History of the Billionaires Behind the Rise of the Radical Right* (New York: Doubleday, 2016), 151.
10. David Rosner and Gerald Markowitz, "Standing Up to the Lead Industry: An Interview with Herbert Needleman," *Public Health Reports* 120 (2005): 330–37.
11. Mitsunobu Tatsumoto and Clair C. Patterson, "Concentrations of Common Lead in Some Atlantic and Mediterranean Waters and in Snow," *Nature* 199 (1963): 350–52.
12. Lucas Reilly, "The Most Important Scientist You've Never Heard Of," *Mental Floss*, May 17, 2017.
13. Justin Scheck, Eliot Brown, and Ben Foldy, "Environmental Investing Frenzy Stretches Meaning of 'Green,'" *Wall Street Journal*, June 24, 2021.
14. Reilly, "The Most Important Scientist You've Never Heard Of."
15. Clare Howard, "Pest Control: Syngenta's Secret Campaign to Discredit Atrazine's Critics," *100 Reporters*, June 17, 2013.

16. Howard, "Pest Control: Syngenta's Secret Campaign to Discredit Atrazine's Critics."
17. Carey Gillam, *Whitewash: The Story of a Weed Killer, Cancer, and the Corruption of Science* (Washington, DC: Island Press, 2017).
18. Matthew Kassel, "Homo Correctus: Paul Brodeur Sues 'American Hustle' to Set the Record Straight," *The Observer,* November 11, 2014.
19. Lauren Kelley, "Author Jane Mayer on How the Koch Brothers Have Changed America," *Rolling Stone,* February 14, 2016.
20. Herbert L. Needleman, "Salem Comes to the National Institutes of Health: Notes from Inside the Crucible of Scientific Integrity," *Pediatrics* 90 (1992): 977–81.
21. John Bach, "Anti-Smoking Crusader," *UC Magazine,* May 2009.
22. Jon Wiener, "Big Tobacco and the Historians," *Nation,* February 25, 2010.
23. Gillam, *Whitewash.*
24. "Freedom to Bully: How Laws Intended to Free Information Are Used to Harass Researchers," Union of Concerned Scientists, February 11, 2015.
25. Wiener, "Big Tobacco and the Historians."
26. Arthur L. Kellermann et al., "Gun Ownership as a Risk Factor for Homicide in the Home," *New England Journal of Medicine* 329 (1993): 1084–91.
27. "How the NRA Suppressed Gun Violence Research," *Union of Concerned Scientists,* October 12, 2017.
28. L. Graves, "Sofia Gatica, Argentine Activist, Faced Anonymous Death Threats for Fighting Monsanto Herbicide," *Huffington Post,* May 3, 2012.
29. "Defending Tomorrow," Global Witness, 2020, https://www.globalwitness.org/.
30. Anne Landman and Stanton A. Glantz, "Tobacco Industry Efforts to Undermine Policy-Relevant Research," *American Journal of Public Health* 99 (2009): 45–48.
31. Rosner and Markowitz, "Standing Up to the Lead Industry: An Interview with Herbert Needleman."
32. Steve Volk, "Was a USDA Scientist Muzzled Because of His Bee Research?," *Washington Post,* March 2, 2016.
33. Suzanne Goldenberg, "Rockefeller Family Tried and Failed to Get ExxonMobil to Accept Climate Change," *The Guardian,* March 27, 2015.

34. Catie Edmondson, "Coll Rebukes Allegations of Misconduct by Columbia Journalists," *Columbia Spectator,* August 10, 2016.
35. Gillam, *Whitewash.*
36. Sara Jerving, "Syngenta's Paid Third Party Pundits Spin the 'News' on Atrazine," *PRWatch,* February 7, 2012.
37. Ray Hilborn and Robert Kearney, "Australian Seafood Consumers Misled by Prophets of Doom and Gloom," 2012, https://thefishsite.com/articles/australian-seafood-consumers-misled-by-prophets-of-doom-and-gloom.
38. Ray Hilborn, "Apocalypse Forestalled: Why All the World's Fisheries Aren't Collapsing," 2010, http://www.atsea.org/doc/Hilborn%202010%20Science%20Chronicles%202010-11-1.pdf.
39. "Centuries-Old Smoking/Health Controversy Continues," January 1968, R.J. Reynolds Records, Master Settlement Agreement, https://www.industrydocuments.ucsf.edu/docs/hgxf0083.
40. Josh Bloom, "BPA Alarmists Are Beating a Dead Monomer—Enough Already," *American Council on Science and Health,* December 30, 2019.
41. "More of the Same: Another Slanted Anti-Pharma Op-Ed. Yawn," American Council on Science and Health, February 12, 2013.
42. Joe Sandler Clarke, "'I Don't Appreciate Them Calling Me a Liar': Bees Study Hits Back at Bayer and Syngenta," *Unearthed,* November 9, 2017.
43. Reilly, "The Most Important Scientist You've Never Heard Of."
44. Elizabeth M. Whelan, *Toxic Terror* (Ottawa: Illinois Jameson Books, 1985), 15.
45. "More of the Same: Another Slanted Anti-Pharma Op-Ed. Yawn," American Council on Science and Health.
46. Freese, *Industrial-Strength Denial.*
47. Cliff White, "Ray Hilborn Fights Back Against Greenpeace Accusations," SeafoodSource, July 14, 2016. https://www.seafoodsource.com/news/supply-trade/ray-hilborn-fights-back-against-greenpeace-accusations.
48. Chris Lischewski, Shue Wing Chan, and In-Soo Cho, "Greenpeace vs. the Tuna Sandwich," *Wall Street Journal,* November 8, 2011.
49. Thomas H. Jukes, "DDT, Human Health and the Environment," *Environmental Affairs* 1 (1971): 3.
50. Roland C. Clement, "The Pesticides Controversy," *Boston College Environmental Affairs Law Review* 2 (1972): 445–68.
51. Gillam, *Whitewash,* 127.

52. Dashka Slater, "The Frog of War," *Mother Jones,* January/February 2012.
53. Nanette Asimov, "UCSF Agrees to $150K Settlement over Sexual Harassment Claim," *San Francisco Chronicle,* October 19, 2018; Dave Cross, "Glantz Settles Claim," *Planet of the Vapes,* October 8, 2018; Jim McDonald, "Stanton Glantz Accused Again of Sexual and Academic Misconduct," *Vaping360,* April 9, 2018.

CASE: STALLING THE PRESERVATION OF ANTIBIOTICS FOR MEDICAL TREATMENT ACT

1. "Summary Report on Antimicrobials Sold or Distributed for Use in Food-Producing Animals," Food and Drug Administration, 2014, http://www.fda.gov/.
2. Consumers Union, letter to Congress, July 31, 2012, https://tinyurl.com/53ahjmnu.
3. H.R. 1587—115th Congress: Preservation of Antibiotics for Medical Treatment Act of 2017," www.GovTrack.us, 2017, https://www.govtrack.us/congress/bills/115/hr1587.

8. CHALLENGE THE POLICY

1. Jennifer Jacquet and Nicolas Delon, "The Values Behind Calculating the Value of Trophy Hunting," *Conservation Biology* 30 (2016): 910–11.
2. Kristin Shrader-Frechette, "Review of *Risk and Reason,*" *Notre Dame Philosophical Reviews,* April 9, 2003.
3. "Concealing Their Sources—Who Funds Europe's Climate Change Deniers?," Corporate Europe Observatory, December 2010.
4. Bill Sells, "What Asbestos Taught Me About Managing Risk," *Harvard Business Review,* March–April 1994.
5. Charles Duhigg, "Debating How Much Weed Killer Is Safe in Your Water Glass," *New York Times,* August 22, 2009.
6. Ryan Koronowski and Joe Romm, "Exxon CEO: 'What Good Is It to Save the Planet If Humanity Suffers?'" *ThinkProgress,* May 30, 2013.
7. Testimony of Mario Lewis, Jr., Vice President for Policy and Coalitions, Competitive Enterprise Institute, before the Committee on Small Business, June 4, 1998, Hearing on the Kyoto Protocol.
8. Testimony of Mario Lewis, Jr., Vice President for Policy and Coalitions, Competitive Enterprise Institute.

9. Anne Landman and Stanton A. Glantz, "Tobacco Industry Efforts to Undermine Policy-Relevant Research," *American Journal of Public Health* 99 (2009): 45–48.
10. Jeff Eshelman, "Divestment Won't Help Environment, but It Could Hurt Pensioners," *Boston Globe,* December 7, 2017.
11. Carey Gillam, "USDA Drops Plan to Test for Monsanto Weed Killer in Food," *U.S. Right to Know,* March 24, 2017.
12. Carey Gillam, *Whitewash: The Story of a Weed Killer, Cancer, and the Corruption of Science* (Washington, DC: Island Press, 2017).
13. Gerald Markowitz and David Rosner, *Deceit and Denial: The Deadly Politics of Industrial Pollution* (Berkeley: University of California Press, 2002).
14. Testimony of Mario Lewis, Jr., Vice President for Policy and Coalitions, Competitive Enterprise Institute.
15. Kelly D. Brownell and Kenny E. Warner, "The Perils of Ignoring History: Big Tobacco Played Dirty and Millions Died. How Similar Is Big Food?," *Milbank Quarterly* 87 (2009): 259–94.
16. Testimony of Mario Lewis, Jr., Vice President for Policy and Coalitions, Competitive Enterprise Institute.
17. Peter Henderson, "Scientists Urge End to Limits on Gun Safety Research," Reuters, January 10, 2013.
18. Molly Peterson, "SoCalGas Admits Funding 'Front' Group in Fight for Its Future," KQED, July 31, 2019.
19. Sylvia Wright, "Don't Blame Cows for Climate Change," UC Davis press release, December 7, 2009.
20. Monsanto Company and nine professors, email exchanges, August 2013, catalogued by *Mother Jones,* https://www.motherjones.com/files/monsanto-sachs-juma.pdf.
21. Ray Hilborn, "Of Mice, Fishermen, and Food," *ICES Journal of Marine Science* 73 (2016): 2167–73.
22. Testimony of Mario Lewis, Jr., Vice President for Policy and Coalitions, Competitive Enterprise Institute.
23. Editorial, "Lead Balloon," *Wall Street Journal,* August 17, 1982, 16.
24. Suzanne Goldenberg, "The Truth Behind Peabody's Campaign to Rebrand Coal as a Poverty Cure," *Guardian,* May 19, 2015.
25. ExxonMobil, "Outlook for Energy: A Perspective to 2040," 2019, https://corporate.exxonmobil.com/.
26. Anne Weir Schechinger and Craig Cox. *Feeding the World: Think US Agriculture Will End World Hunger? Think Again,* 2016, https://

static.ewg.org/reports/2016/feeding_the_world/EWG_FeedingThe World.pdf?_ga=2.266030092.222165887.1639065189-2080776524 .1639065189.

27. Naomi Oreskes and Erik M. Conway, *Merchants of Doubt: How a Handful of Scientists Obscured the Truth on Issues from Tobacco Smoke to Global Warming* (London: Bloomsbury, 2010).
28. Barbara Freese, *Industrial-Strength Denial: Eight Stories of Corporations Defending the Indefensible, from the Slave Trade to Climate Change* (Berkeley: University of California Press, 2020).
29. Robert Stavins, "Pitching Divestment as a 'Moral Crusade,'" *New York Times,* August 10, 2015.
30. Stavins, "Pitching Divestment as a 'Moral Crusade.'"
31. Ray Hilborn, "Marine Protected Areas Miss the Boat," *Science* 350 (2015): 1326.
32. Laura Nixon et al., "'We're Part of the Solution': Evolution of the Food and Beverage Industry's Framing of Obesity Concerns Between 2000 and 2012," *American Journal of Public Health* 105 (2015): 2228–36.
33. Herbert L. Needleman, "Salem Comes to the National Institutes of Health: Notes from Inside the Crucible of Scientific Integrity," *Pediatrics* 90 (1992): 977–81.
34. Markowitz and Rosner, *Deceit and Denial.*
35. Sean Paige, "Zoned to Extinction," *Reason Magazine,* October 2001.
36. "Global Warming Petition Project," http://www.petitionproject.org/.
37. Ray Hilborn, "Environmental Cost of Conservation Victories," *Proceedings of the National Academy of Sciences* 110 (2013): 9187.
38. Will Flanders, "This Is How Devastating the Green New Deal Would Be for Wisconsin," *Hill,* March 14, 2020.
39. Testimony of Mario Lewis, Jr., Vice President for Policy and Coalitions, Competitive Enterprise Institute.
40. Thomas Whiteside, "A Cloud of Smoke," *New Yorker,* November 30, 1963.
41. Jim McDonald, "Meet the Rich Moms Who Want to Ban Vaping," *Vaping360,* October 8, 2018.
42. Nixon et al., "'We're Part of the Solution': Evolution of the Food and Beverage Industry's Framing of Obesity Concerns Between 2000 and 2012."
43. Alaska Delegation Letter to the Federal Reserve, OCC, and FDIC, June 16, 2020, https://www.documentcloud.org/documents/6950908 -06-16-20-AK-Delegation-Letter-to-the-Federal.html.

44. Emily Atkin, "The Quiet Campaign to Make Clean Energy Racist," *Heated* (newsletter), July 2, 2020.
45. Nixon et al., "'We're Part of the Solution': Evolution of the Food and Beverage Industry's Framing of Obesity Concerns Between 2000 and 2012."
46. Stanley Fish, "Divesting from Fossil Fuels: The Student Assault on the Academy," *Huffington Post,* November 17, 2016.

CASE: HUMAN NATURE CAN HELP JUSTIFY INACTION

1. Jonathan Franzen, "What if We Stopped Pretending?," *New Yorker,* September 8, 2019.
2. Quentin Atkinson and Jennifer Jacquet, "Challenging the Idea That Humans Are Not Designed to Solve Climate Change," *Perspectives on Psychological Science* (2021): doi.org/10.1177/17456916211018454.

9. OUTSIDE OPPORTUNITIES

1. Nina Mazar and Daniel Ariely, "Dishonesty in Scientific Research," *The Journal of Clinical Investigation* 125 (2015): 3993–96.
2. Quentin Atkinson and Jennifer Jacquet, "Challenging the Idea That Humans Are Not Designed to Solve Climate Change," *Perspectives on Psychological Science* (2021): doi.org/10.1177/17456916211018454.
3. Mark Van Vugt, Vladas Griskevicius, and P. Wesley Schultz, "Naturally Green: Harnessing Stone Age Psychological Biases to Foster Environmental Behavior," *Social Issues and Policy Review* 8 (2014): 1–32.
4. George Marshall, *Don't Even Think About It: Why Our Brains Are Wired to Ignore Climate Change* (London: Bloomsbury, 2014).
5. Atkinson and Jacquet, "Challenging the Idea That Humans Are Not Designed to Solve Climate Change."
6. "Forwarding Memorandum," May 19, 2008, Philip Morris Records, Master Settlement Agreement, https://www.industrydocuments.ucsf.edu/docs/ggkj0191.
7. Nathaniel Rich, "Losing Earth: The Decade We Almost Stopped Climate Change. A Tragedy in Two Acts," *New York Times Magazine,* August 5, 2018.
8. Jennifer Jacquet, *Is Shame Necessary? New Uses for an Old Tool* (New York: Pantheon, 2015).

9. Anand Giridharadas, *Winners Take All: The Elite Charade of Changing the World* (New York: Alfred A. Knopf, 2018), 97.
10. Jon Wiener, "Big Tobacco and the Historians," *Nation,* February 25, 2010.
11. Karen Miller, *The Voice of Business: Hill & Knowlton and Postwar Public Relations* (Chapel Hill: University of North Carolina Press, 1999).
12. Susanne C. Moser and Lisa Dilling, eds., *Creating a Climate for Change: Communicating Climate Change and Facilitating Social Change* (Cambridge: Cambridge University Press, 2007), 4.
13. Astrid Zweynert, "Stop 'Boring' Language to Spur Climate Action, U.N. Environment Chief Says," Thomson Reuters Foundation News, December 19, 2017.
14. Marc Parry, "Princeton Climate Skeptics Tried to Ignore a Campus Skeptic. Then He Went to the White House," *Chronicle of Higher Education,* August 16, 2019.
15. Warren Pearce, "We'll Never Tackle Climate Change If Academics Keep the Focus on Consensus," *Guardian,* August 1, 2017.
16. Barbara Freese, *Industrial-Strength Denial: Eight Stories of Corporations Defending the Indefensible, from the Slave Trade to Climate Change* (Berkeley: University of California Press, 2020).
17. Giridharadas, *Winners Take All.*

CASE: GROWING DISTRUST IN BIG BUSINESS

1. "Edelman Trust Barometer 2020," Edelman, 2020.
2. Ed Williams, "The UK: A Parable of Distrust," Edelman, January 19, 2020.
3. "A Roadmap for Stakeholder Capitalism: 2019 Survey Results," JUST Capital.
4. "Trust and Mistrust in Americans' Views of Scientific Experts," Pew Research Center, August 2019.

10. NEAR-TERM THREATS

1. Bill Sells, "What Asbestos Taught Me About Managing Risk," *Harvard Business Review,* March–April 1994.
2. Marc Gunter, "Edelman Loses Executives and Clients over Climate Change Stance," *Guardian,* July 7, 2015.

3. Jay Greene, "Amazon Employees Launch Mass Defiance of Company Communications Policy in Support of Colleagues," *Washington Post,* January 27, 2020.
4. "HP Workforce Sustainability Survey Global Insights Report," Edelman Intelligence, April 2019.
5. Sisi Wei, Annie Waldman, and David Armstrong, "Dollars for Profs," *ProPublica,* December 6, 2019.
6. Paulo M. Serôdio, Martin McKee, and David Stuckler, "Coca-Cola—A Model of Transparency in Research Partnerships? A Network Analysis of Coca-Cola's Research Funding (2008–2016)," *Public Health Nutrition* 21 (2018): 1594–1607.
7. Lisa Kearns and Arthur Caplan, " 'A Little ELF, Please?' The Electronic Long-Form COI Disclosure Statement (ELFCOI)," *The American Journal of Bioethics* 18 (2018): 1–2.
8. Hiroko Tabuchi, "How Climate Change Deniers Rise to the Top in Google Searches," *New York Times,* December 29, 2017.
9. Leo Hickman, "Do Critics of UN Meat Report Have a Beef with Transparency," *The Guardian,* March 24, 2010.
10. Ed Yong, "Wait, So How Much of the Ocean Is Actually Fished?," *The Atlantic,* September 10, 2018.
11. "Foreign Secretary Boosts BBC Funding to Fight Fake News," May 1, 2021, https://www.gov.uk.
12. Paul Brodeur, "The Asbestos Industry on Trial," *New Yorker,* June 17, 1985.
13. Elliott Negin, "ExxonMobil Claims Shift on Climate but Continues to Fund Climate Science Deniers," *Union of Concerned Scientists/The Equation,* October 22, 2020. Also see Chris McGreal, "ExxonMobil Lobbyists Filmed Saying Oil Giant's Support for Carbon Tax a PR Ploy," *Guardian,* June 30, 2021.
14. Eric. M. Meslin et al., "Benchmarks for Ethically Credible Partnerships Between Industry and Academic Health Centers: Beyond Disclosure of Financial Conflicts of Interest," *Clinical and Translational Medicine* 4 (2015): 36.
15. Jennifer Rubin, "Time to Confront the Right-Wing Myth-Makers," *Washington Post,* November 6, 2020.
16. David Dayen, "Corporate-Funded Judicial Boot Camp Made Sitting Federal Judges More Conservative," *Intercept,* October 23, 2018.
17. Elliott Ash, Daniel L. Chen, and Suresh Naidu, working paper, "Ideas Have Consequences: The Impact of Law and Economics on

American Justice," http://users.nber.org/~dlchen/papers/Ideas_Have_Consequences.pdf.

18. David Garcia, Victor Galaz, and Stefan Daume, "EATLancet vs. Yes2meat: The Digital Backlash to the Planetary Health Diet," *The Lancet* 394 (2019): 2153–54.